Weight and Weight

Question:
I heard someone talk about weightlessness as one of the problems of space flight. I think this means that things would have no weight. But how can this be? I would think that if something didn't have any weight, there wouldn't be any of it at all.

Yes, how can there be weightlessness? It would mean that you could hold a rubber ball in your hand, feel it, squeeze it, know that it was really there, and yet it would not weigh anything. If you put it in front of your nose and let go, it would just stay there. And maybe if this were to happen to you, you wouldn't weigh anything either. But on the surface of the earth where we live, this doesn't happen, does it? We are sure of the answer to that one from our own experience.

Weightlessness is, sure enough, one of the problems of space flight. But first we had better talk about what weight is. And since weight depends on two things, we had better talk about each of these.

Suppose that you have in your hand a solid rubber ball. How much stuff, how much rubber, is in the ball? Its volume — how big it is — is not a very reliable measure. We could squeeze it and make it smaller without changing how much rubber is in the ball. So we invent a name for how much stuff there is in the rubber ball, or in any object. We call it mass. The idea is that mass is a measure of how much. And for any object, like the rubber ball, its mass is always the same, no matter where it is.

Mass has an important quality. Any two objects have a pull or force which attracts them to each other. It is as if every object were at least a little bit sticky toward every other object even when they are apart. The force or pull or stickiness between them depends upon the masses of the two objects and how close they are together. This is called gravitation.

The stickiness or pull between most objects of our experience really is pretty small. If both objects are little, or even so big that they weigh hundreds of pounds each, the pull between them

is very small. For example, it would be difficult to measure the very small force of attraction between two automobiles. (It is not gravitation which causes cars to bump into each other!) But if the objects are very large, or if even just one of them is very, very

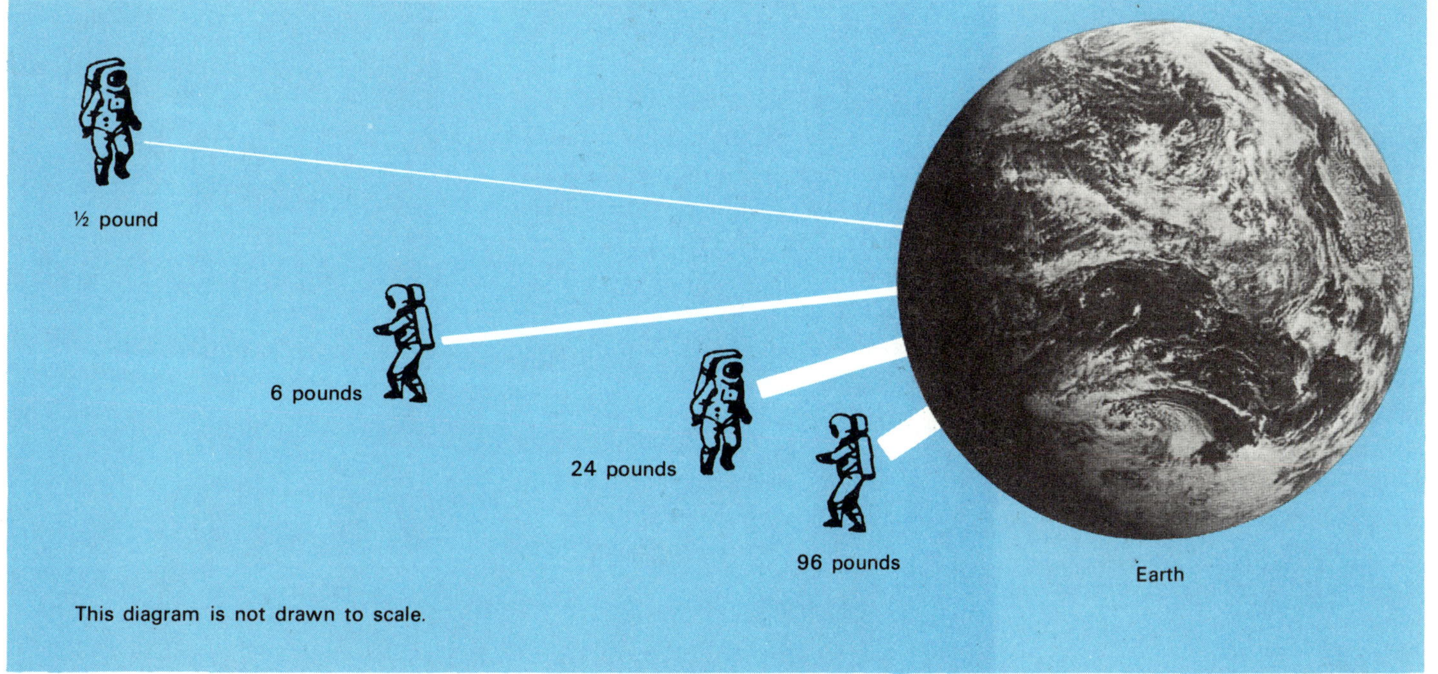

This diagram is not drawn to scale.

large, then the force between them can be so great that it is easy to measure.

Now, there is one very, very large object which you touch every day. Have you guessed it? It's the earth. Because the earth has so very much mass, there is a large force of attraction between it and even rather small objects. This force, between the earth and objects near it, we call gravity. You knew about this all the time, didn't you? We all learned about gravity the hard way when we were very small kids — and fell down.

But I'll bet there are some things about gravity that you have not thought about. It allows us to measure the mass of anything on or near the earth's surface. If you say that you weigh 96 pounds, what you really mean is that you have an amount of mass which is always pulled toward the earth with a force of 96 pounds. You also mean that you are pulling on the earth with a force of 96 pounds!

There is one more thing about gravitation we need to know. The force of attraction of any object always works as if all of its mass were right at its center. So really the force of gravity tends to pull things toward the center of the earth. The earth is almost round, and the distance to the center almost the same everywhere. So any given mass (like your body) weighs just about the same, anywhere you go. And that's mighty convenient.

But maybe you know that the earth is not perfectly round. It's just a little bit flattened or squashed down at the north and south poles. This means that as you travel north from New Orleans, you weigh more — unless you climb a mountain. You are getting a little closer to the center of the earth. If you weigh 96 pounds in New Orleans and should take a very quick trip to Chicago, you would gain in weight by about an ounce and a half.

Now that we know about weight, what about weightlessness? You can see from the diagram how much you might weigh if you could stop somewhere out in space. You would always weigh something. Even at 12,000 miles out the earth would be pulling you back.

As a practical matter you could not very well just stop somewhere out in space. However, there is a special kind of motion which would keep you out there.

Suppose you are in a satellite capsule circling the earth. Then your very movement makes you weightless. Your motion gives a force tending to whirl you out away from the earth. And this is just as great as the force of gravity tending to pull you in. There is still just as much of you. You have the same mass. But now you are weightless.

Try This!

Use a Piece of Glass To Copy Pictures

Buy a pane of window glass at your hardware store. A small piece about 8 by 10 inches will do very well. Put tape over all four edges so you won't cut yourself while handling the glass.

Place on a table the picture you want to copy. Stand the glass along one edge of the picture. On the other side of the glass put a piece of writing paper. Look through the glass at the writing paper. You will seem to see the picture there on the writing paper. Now you can trace around the picture, making a copy of it on the writing paper. But notice that everything is **backwards,** just like an image seen in a mirror.

How It Works

When a beam of light strikes glass, most of it travels through the glass, but some of it bounces off the surface and is **reflected.**

When you look through the glass, you actually see the writing paper but you also see a reflection of the picture on the glass. To your eye, the reflected picture seems to be on the writing paper.

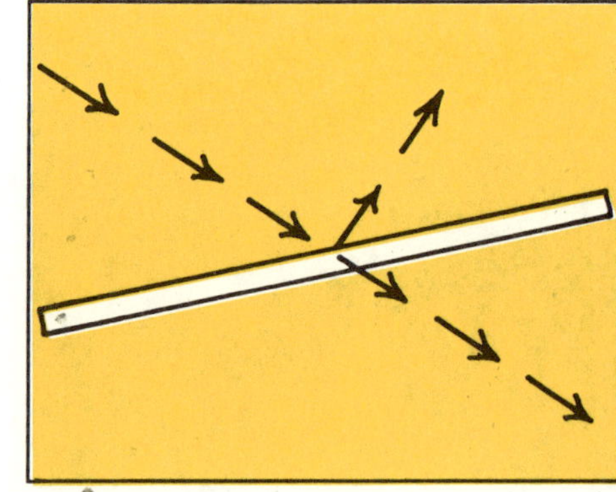

Why It Weighs Less in Water

Question:
Last summer I went swimming in a creek on my grandfather's farm. I was building a dam out of rocks. I found one rock underwater which I could lift up to the top of the water. But then it seemed to get heavier. I never could lift it all the way out of the water. Why was this?

I guess we have all had an experience something like yours. Let's start by thinking about the more common experience (I hope) of getting into a bathtub. The next time you take a bath, fill the tub about half full. Then mark where the water level is, maybe with a piece of adhesive tape. After you get in the bathtub, check where the water level is. It's higher. Why?

I'm sure you know the answer. You and the water can't occupy the same space at the same time. When you went in, some of the water had to move. And where could it go? It just had to go up and it raised the water level in the tub. If you go swimming in the ocean you also raise the water level a little. But the ocean is so big that the rise would be hard to measure.

Now, this is the basic idea. When any object is submerged in a fluid such as water, it pushes away some of the fluid. Since the fluid is trying to get back, there is an upward push on the object. The upward push is called a buoyant (boy-unt) force. And we can always figure what the buoyant force is. It is equal to the weight of the fluid which the object pushes away or displaces.

Suppose that water is our fluid. A pint of water weighs just about one pound. Now a pint of lead weighs about 12 pounds. Suppose we drop a pint of lead into water. There is still the same mass (amount) of lead there. And it is still pulled down by gravity by a force of 12 pounds. But there is a pint or a pound of water trying to get back where the lead is. There is a buoyant force of one pound pushing upward on the lead. So, underwater, a pint of lead would weigh only 11 pounds.

Let's think about your rock underwater. Suppose you can lift a weight of 50 pounds. Now, 50 pounds of rock would have a volume of about 20 pints, depending on the kind of rock. Underwater, the buoyant force of water makes the rock 20 pounds lighter, so that it weighs only 30 pounds. But when you begin to lift it out of water, it seems to suddenly change in weight from 30 pounds to 50 pounds.

Not all kinds of things have the same mass or weight of stuff in a given volume. This property of things we call density. In the table on the next page, you will see the densities of several kinds of material compared to water. To determine the density of an object, we need two measurements. First we have to weigh it. Then we have to measure its volume.

An easy way to measure the

volume, even for an irregular object, is to drop it in a pitcher full of water and see how much water spills over. For example, 150 pennies (made mostly of copper) weigh just about one pound. Look up the density of copper in the table and see if you can figure out how much water would spill over from a pitcher full of water if you dropped 150 pennies into it.

The idea of the buoyant force of a fluid has been known for a very long time. The first person to really figure it out was Archimedes (ark-ih-ME-deez) who lived in Sicily about 250 years before the birth of Christ. And there is an interesting story about his discovery.

Archimedes was a good friend of the king. The king had a crown made for him which was supposed to be solid gold. But he suspected that the goldsmith who made it had cheated him by using silver and just covering it with gold. So the king asked Archimedes, "How can we find out if the crown is pure gold without cutting it up?"

Archimedes puzzled over this for a long time. One day he was taking a bath. And he made the same observation that I asked you to make about the water you displace when you get into a bathtub. All of a sudden the idea came to him. He had discovered how to use the buoyant force of water. And he was so excited that he jumped out of his bath and took off for the palace without his clothes, shouting "Eureka!" which meant "I have found it!"

Archimedes tested the crown. He compared it to pure gold and to pure silver. He found that the goldsmith had been honest. The crown was really pure gold.

And now my question to you is: Just how did Archimedes go about showing that the crown was pure gold and not silver with a little gold on the outside?

If you cannot figure out just what Archimedes did, write to me and let me explain. But first try hard to figure it out yourself. You will need some of the information in the table.

The Density of Some Things

	Pounds per pint
Water	1
Rock	2½
Copper	9
Silver	11
Lead	12
Gold	20

(For example, a pint bottle of water weighs about 1 pound, and a pint of gold weighs 20 pounds.)

Try This!

Here's something you can try almost any old time — maybe when you are waiting for your folks to read your report card.

With scissors, cut a square out of a piece of thin cardboard. The cover from a used pack of paper matches will do. The size is not important, but first try one about 1 inch on each side.

Bend one corner toward you. Bend the opposite corner away from you. Now hold the square by pressing lightly between your thumb and finger against the unbent corners. You have a cardboard turbine.

Blow against it and you can make it spin. You must hold it with just the right tightness. If your fingers are too tight, it won't spin. If your fingers are too loose, the card will blow away. Blow against the side rather than the center. If you hold it right and blow hard, you can really rev it up.

If you want to get fancier, you can make a moving picture. Use a piece of white cardboard to make your turbine. Mark an X in the center on one side. Mark an O in the center of the other side. Now make it spin. When it is spinning fast, you will be able to see both the O and an X inside of it.

How It Works

You can see how your turbine works. The corners are turned so that wind always pushes the card around in the same direction. Turn the card upside down and see what happens to the direction of rotation.

If the card turns fast enough, it fools your eye just as a movie does. Your eye "remembers" each picture for a short period of time. If pictures come fast enough, they add together as the O and the X did on your card.

Will It Sink or Float?

Question:
Why do some things float on water when others do not? Wood will float but steel will not. I can see how you can build a ship of wood. But how can a ship float when it is built of steel?

We have talked before about the buoyant (boy-unt) force of water. When we put something like a rock in water, the rock must push away or displace its own volume of water. The rock sinks. But it seems lighter in water than it did in air. Since the water is trying to get where the rock is, there is an upward push or buoyant force.

Now suppose that instead of a rock we use a piece of wood. We say that the wood floats. But what does this mean? Put a piece of wood in a glass or a pan of water. You will see that part of the wood sinks below the surface. The wood sinks until it displaces its own weight of water. So if an object weighs less than an equal volume of water, it will float partly above the surface. If an object weighs more than an equal volume of water, it will sink.

You can do an experiment to learn about floating bodies with a glass of water, a cork, and an ice cube. A cork weighs about ¼ as much as an equal volume of water. So when you drop a cork into a glass of water, it will float with about ¼ of it below the water surface. Now put an ice cube in the glass of water. It floats. But notice that most of it is below the water surface. Ice weighs about 9/10 as much as an equal volume of water. So an ice cube sinks until it displaces 9/10 of its volume of water and floats with only about 1/10 of the cube above the surface.

Now how about the ship that floats even though it is made of steel? Steel is almost 8 times as heavy as its own volume of water. So a piece of steel always sinks. But does it? Let's do an experiment to find out. All we need is a metal cap from a soft-drink bottle. Except for the little inside liner, this is a thin piece of iron. And iron is about the same density as steel.

Does a bottle cap float or sink in a glass of water? The answer depends upon how you put it on the water. What happens when you put it edge first into the water? What happens when you hold it carefully upside down and put it onto the water? Why does it float one way and not the other?

The answer is the same for all floating bodies. An object floats if it displaces more than its own weight of water.

Doing these simple experiments with a bottle cap tells you a lot about floating bodies. Even steel can be made to float. The trick is to give it a shape which makes it displace more than its own weight of water.

Air Has Weight, Too

Question:
I have seen balloons at a circus which seemed to stand straight up on a string. One boy let his go and it just kept rising until the wind carried it away. How can you make a balloon rise? When I blow one up with air, it always sinks slowly to the ground.

The answer to this goes way back to Archimedes who sat in his bathtub, watched the water rise, and got a wonderful new idea. It depends upon the **buoyant force** of air like the buoyant force of water which we talked about before. Let's begin by talking about air. Air is pretty common and there is lots of it around us. But it really is interesting stuff.

The first idea is that air really is something. You have felt the wind blow against you. So there must be something around you, even though you can't see it. If you fill a soda pop bottle with water and blow into it with a straw, some of the water will spill out of the bottle. So air takes up space. In water, bubbles of air rise rapidly to the surface. This proves that a bubble of air is lighter than an equal volume of water.

Does air weigh anything at all? It most certainly does — though not very much. A quart of air weighs about 4/100 of an ounce. How do we know this? Well, suppose we had a light but very strong bottle which held just a quart and which had a valve on top that could be opened or tightly closed. We could weigh it (with air in it) on a very sensitive scale. Then we could pump all the air out and close the valve. With all the air pumped out, the quart bottle would weigh 4/100 of an ounce less than it did when filled with air.

Air is a **fluid.** It can move and flow even easier than water. So living with air all around and above you is a little like being at the bottom of a very large swimming pool. Remember that, in water, an object like your body seems to weigh less than it does in air. The water is trying to get down where your body is. And if your body is lighter than an equal volume of water, you can float.

Now just the same thing happens in air except that your body weighs very much more than an equal volume of air. So you can't float in air (in case you haven't tried). But there is a buoyant force of air and you are really just a little heavier than any scale can tell you. If you weigh 100 pounds, you have a volume of 50 quarts. There are 50 quarts of air trying to get where you are, and this much air weighs about 2 ounces. So you are about 2 ounces lighter than if there were no air around you at all.

How can we make something float or rise in air? You will see that this is the same kind of problem as finding something which will float in water. Something will rise in air if it weighs less than an equal volume of air. Can we find anything that weighs less than air?

Air is a mixture of gases, mostly **nitrogen** and **oxygen.** Now there are a number of other gases even lighter than air. The lightest is **hydrogen,** and the next lightest is **helium.** If we had a 1,000-quart tank, it would hold about 40 ounces of air, but only 6 ounces of helium and only 3 ounces of hydrogen. So you will see that if we fill a balloon with hydrogen or helium (and if the rubber of the balloon is not too heavy), then the whole balloon weighs less than an equal volume of air. And so it rises. And if we make the balloon big enough, it can lift a person. The Air Force once built one big enough to lift a man in a sealed cabin up to 100,000 feet above the earth.

Actually hydrogen is a dangerous gas to use because it can be explosive. And helium is pretty hard to get these days because there isn't much available, and it is needed for medical and chemical work. But whenever you see a toy balloon stand up straight on a string, or rise and float away, you can be sure it is filled with some gas which is lighter than air.

Try This!

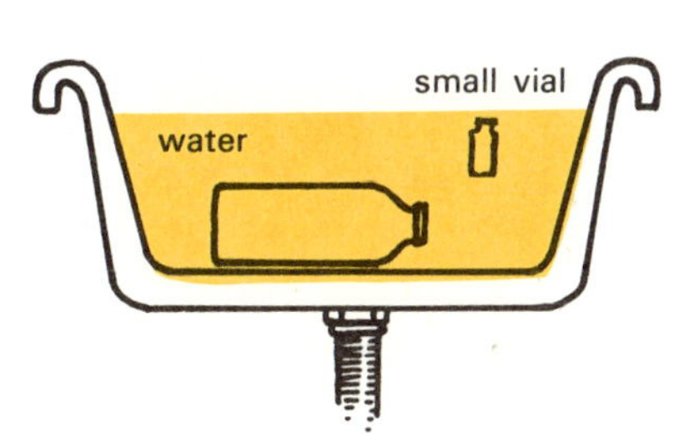

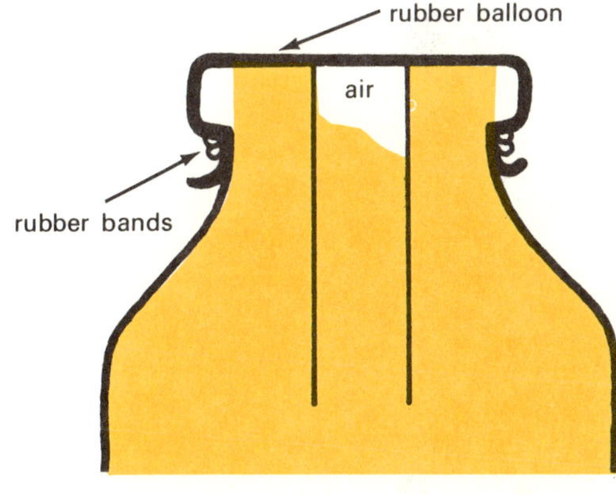

You will need a milk bottle; the top from a lipstick tube or a small glass vial of similar size (small enough to fit through the neck of the bottle); a toy balloon; and a few rubber bands.

Fill your kitchen sink about half-full of water. Lay the milk bottle on its side in the sink so that it is filled with water. Now you must experiment a little to fix the diver (the lipstick tube or vial) so that it will just barely float. Fill it about three-quarters full of water, put your thumb over the open end, and put it beneath the surface, upside down. Let go. If it sinks, try again with less water. If it floats and sticks up much above the surface, try again with more water. You must get the diver with the right amount of air in it so that it will just barely float.

Now tilt the milk bottle and slip the diver into the bottle. Then carefully turn the bottle right side up with the diver floating in it. You may have to pour a little more water into the bottle to fill it right to the top.

Now stretch the balloon over the top of the bottle and fasten it with rubber bands.

When you press on the rubber cover of the bottle, the diver will sink. When you stop pressing, the diver will rise again. With a little practice you make the diver rise and fall at will, even stop motionless in the middle of the bottle.

And if you don't have a balloon, you can make it work by just pushing down on the open mouth of the bottle with the palm of your hand.

Why Does It Work?

When you push down, you are squeezing all the contents inside the bottle. Air will compress into a smaller volume but water will not. When you press down, you push more water into the diver and compress its air into a smaller space. That makes the diver heavier and it sinks.

When you release your hand, the air in the diver pushes back and pushes some of the water out of the diver. It becomes lighter and rises.

How Heavy Is a Pound?

Question:
You told us once that a person in Chicago would weigh about an ounce more than he would weigh in New Orleans. If this is so, why couldn't I make money by buying gold in New Orleans and selling it in Chicago?

Say, it almost looks as if you have something there. If you did, I would ask for a partnership in your new company. But actually the answer is: No, it won't work. Let me explain why.

Perhaps you remember that the real measure of the amount of stuff in any object is its **mass.** And the mass of a gold bar is the same wherever it is — whether it is in New Orleans or Chicago or out in space. Perhaps you remember, also, that a convenient way to measure the mass of something is to measure its **weight.** And the weight of a body depends upon its mass and on the **force of gravity** which pulls it toward the center of the earth. At different places on the earth's surface the force of gravity is almost, but not quite, the same. At Chicago the force of gravity is just a little bit greater than it is at New Orleans, and so any object weighs just a little more at Chicago.

Now there are two methods by which we can weigh things. One way is to measure the stretch of a spring. The more we pull on a spring, the more it stretches. So we can make a **spring scale.** It really measures the weight of an object, the force with which it is pulled toward the earth. If you got in a spaceship and took your spring scale far out in space, where the pull of gravity was very low, it would show that your object weighed much, much less. So a spring scale really can show that an object weighs more in Chicago than in New Orleans. Unfortunately, most spring scales are not very accurate and would not be able to measure such a small difference.

There is another and more accurate method of weighing. It works on the same principle as the teeter-totter. You probably played on a teeter-totter when you were little. Try to remember how it worked. Suppose that you got on one end and a friend got on the other, and you were both the same distance from the center post. Suppose neither of you pushed but just raised your feet. Who would go up and who would go down? Why, certainly, the heavier one would go down. Or if both of you weighed the same you would just balance each other. We can use this same principle to compare the weights of any two objects. Actually the gadget for doing this doesn't look much like a teeter-totter even though it works on the same idea. We call it a **balance.**

In order to weigh with a balance, we need a set of **weights** — pieces of brass or steel the weights of which we already know. If we want to weigh a bar of gold, we would put it on one side of the balance. Then we would put weights on the other side until neither side went up or down — until both sides were just balanced. Then we would add up the weights we used, and this amount would be the weight of the gold bar.

One advantage of the balance method is that it doesn't make any difference at all where we do the weighing. Suppose we weigh the gold bar in New Orleans. And then we take the gold bar and our weights to Chicago. The bar gets a little heavier, but so do our weights. And even if we

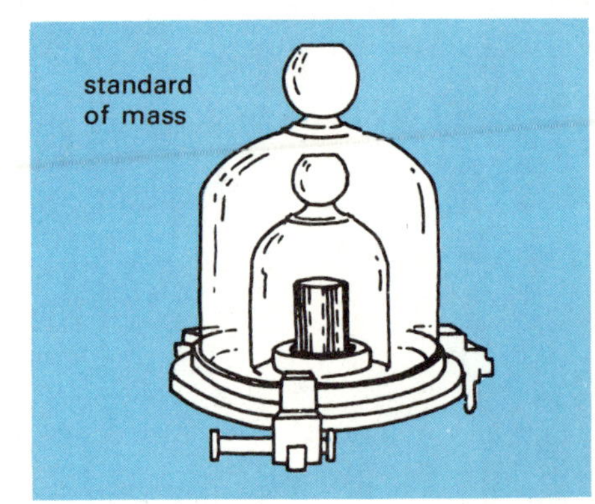

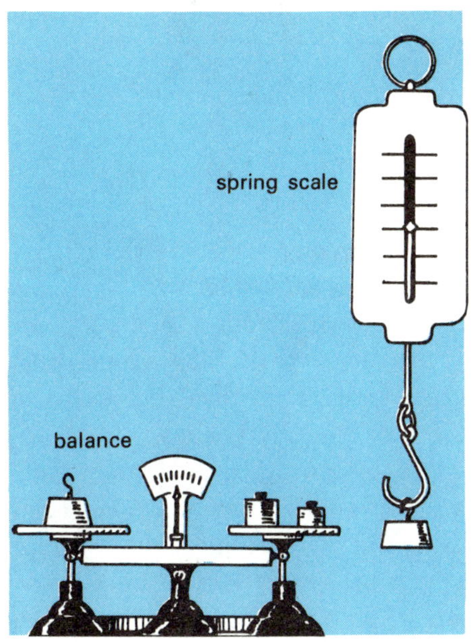

did our weighing on a spaceship far out in space where the force of gravity is very small, exactly the same weights would still balance the bar of gold. Do you see the idea? A balance becomes a way of measuring mass because it is not affected by changes in the force of gravity.

Since gold is so expensive, it is always weighed on a very sensitive balance. So you see why you and I cannot make a fortune buying gold in New Orleans and selling it in Chicago.

You may have thought up another question which we had better talk about right now before I forget. How do we know that the weights which we use on a balance are correct? How do we know that the 1-pound weights used in New Orleans and Chicago and New York all have the same mass? In fact, how do we know that a pound is always a pound? The answer is that somewhere there has to be a weight which everyone agrees on — a standard.

In our National Bureau of Standards in Washington, D.C., there is a cylinder made of two metals, platinum and iridium, which do not rust or corrode. This little cylinder is just exactly like other cylinders carefully kept by other countries as their standards. All of them were made at the same time. Each of them weighs just exactly one **kilogram** which is a metric unit of mass. In the United States we commonly use a smaller unit of mass (which we also call weight), the **pound.** And we say that a kilogram equals 2.2046 (about 2 1/5) pounds.

We also have other standards for other kinds of measurements like **length** and **time,** and even standards to measure the brightness of lamps. Keeping the standards and helping scientists to have a good set of standards for their measurements is one of the important jobs of our National Bureau of Standards.

Try This!

You can make a simple musical instrument, play simple tunes, and learn something about the nature of sound.

All you need is a pail and a string for a guitar or violin. You can buy such a string at a music store. The steel wire string works best.

Fasten the wire securely across the top of the pail, stretching it tightly between the handle loops and tying it so that it cannot slip.

When you pluck the wire with a stick or nail, it will produce a musical note. Put the pail under your arm and squeeze the sides. As you press on the pail, you will stretch the wire still tighter, and it will produce a higher note. With a little practice you can play simple tunes by squeezing the pail as you pluck the wire.

How It Works

The wire produces a musical note because it vibrates rapidly when it is plucked.

This sets up sound waves in the air because every time the wire moves back and forth, it pushes air with it. The air next to the wire pushes the air beyond, and rapidly the "sound" reaches your ear.

The note you hear depends on how fast the string vibrates. If it moves back and forth slowly, it makes a low note. But if it vibrates very fast, it makes a higher note.

How fast a wire vibrates depends on two main things: (1) how long the wire is, and (2) how tight the wire is stretched.

Most stringed instruments change the sound by making the vibrating part of the string longer or shorter. That's what a violin or guitar player does with the fingers of one hand.

But your instrument changes notes by stretching the wire tighter. This is the way your own vocal chords make sounds. Sing a low note and then a high note. Feel your throat tighten as you try for the high note.

Watch a Falling Body

Italian Government Travel Office Photo

Question:
My father says that when things fall to the ground, their speed is just the same, no matter what they weigh. I would think that something heavier would fall faster than something lighter. We tried an experiment to find out. Dad stood up on our kitchen stool and dropped a dime and a half dollar so that they started at the same time. Each time we tried it, the coins seemed to land right together. So it looks as if Dad is right, but it is hard for me to believe. It just seems that a heavy coin should fall faster and reach the floor sooner than a light one. If the speed of something falling does not depend on its weight, what does it depend on?

May I compliment you and your father for trying the experiment. What your father said and what the experiment told you are right. The speed of a falling object is almost independent of its weight. We will get to the "almost" part later.

And you should not feel badly. One of the wisest of men once thought the same thing you did. Some four hundred years before the birth of Christ there lived in Greece a man named Aristotle (pronounced air-iss-TOT-el). He carefully observed the world around him and tried to understand it. He wrote down what he saw and believed. Because he was so wise, people respected what he said. And for the following 2,000 years, people learned what was known about science by learning what Aristotle had taught.

Now, none of us are perfect and we do make mistakes. You should realize that even today's scientists sometimes make mistakes. Even in the textbooks used in universities there can be mistakes. So it is not surprising that Aristotle should have made mistakes, too. But it does seem surprising that for 2,000 years — and that's a mighty long time — people should have learned about science just by memorizing what Aristotle had said. After all, science is learning about the world around us. It is a pretty marvelous world, and memorizing what someone else has said seems a dull way to study science.

One of the mistaken ideas which Aristotle had written down was that the speed of a falling body depends upon its size or weight. I don't know how he got this idea. I'm sure he did not think of doing the experiment which you and your father did with the two coins. And I guess that for 2,000 years no one else did the experiment, either — they just believed what Aristotle had taught.

That's how things were until 1590 (about 400 years ago) when another man realized that he could learn about science for himself. At the University of Pisa in northern Italy, there was a young teacher named Galileo Galilei. We remember him by his first name, Galileo (pronounced gal-i-LAY-oh), as one of the world's great scientists.

Because Galileo really was a scientist, he DID the experiment. In Pisa there is still standing a monument

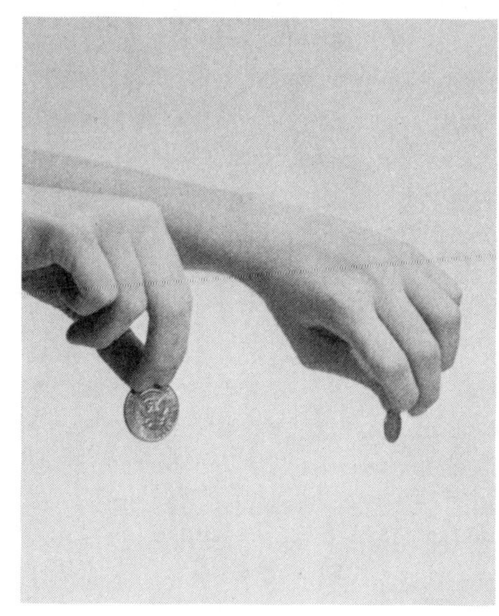

which tips a little, the famous Leaning Tower of Pisa. Galileo tried his experiment by dropping iron balls from the top of the tower. He let go together a big ball weighing about 10 pounds and a small ball weighing about 1 pound. The two balls hit the ground right together. Most people didn't believe what they saw, or maybe they just didn't want to believe. They had learned what Aristotle had taught. And it is upsetting for people to realize that something which they had believed to be true is wrong. Galileo found that he was a mighty unpopular man.

Why should it be that the speed of fall of an object does not depend upon its weight? After all, weight is just the pull on an object by the force of gravity. A heavier weight gets more pull. But it is also true that the bigger and heavier an object is, the more force it takes to get it moving. Try throwing a very big stone and a small stone. Which of them can you throw the faster? I think you will get the idea. A heavier object has more pull on it, due to gravity. But because it is heavier it takes just that much more pull to get it moving. The two effects cancel out each other. And a big stone and a small stone fall to the ground right together.

So, you see, your experiment came out just the same as Galileo's. I guess his experiment was a more careful one because he was able to use a tower almost 180 feet tall. Galileo didn't stop there. Just as you did, he asked other questions about falling bodies. What does determine how fast they fall? But it looks as if I have spent too much time talking about Aristotle and Galileo. In the next article I will try to answer your last question.

Try This!

You can do some important experiments about **weight.** All you need are bathroom scales, a board about 6 feet long which will support your weight, and a couple of friends about your size.

Experiment No. 1

Stand on the scales and see what you weigh.

If you stand on the scales on just one foot, do you weigh half as much as if you stand on two feet? Try it and see.

Experiment No. 2

Put one end of the 6-foot board on the scales, and the other end on a box or stack of magazines about the same height as the scales. Then adjust the pointer on the scales so it reads zero. (Most scales have a little adjusting knob or wheel to turn until the scales read zero when you are not standing on them.)

Now stand on the board right over the scales as in A. Have a friend look at the scales. What do the scales read?

Next, move about halfway down the board to B. Now what do the scales read? Where did your weight go?

Move to position C. Now what do the scales read?

Experiment No. 3

Put the board across the scales like a teeter-totter, and again adjust the pointer to zero. You stand on one end and have a friend stand on the other end, so that you balance. What do the scales read?

Now move together so you are both right over the scales. Does it read any differently? Why? Do the scales read less as you both move away toward the ends of the board?

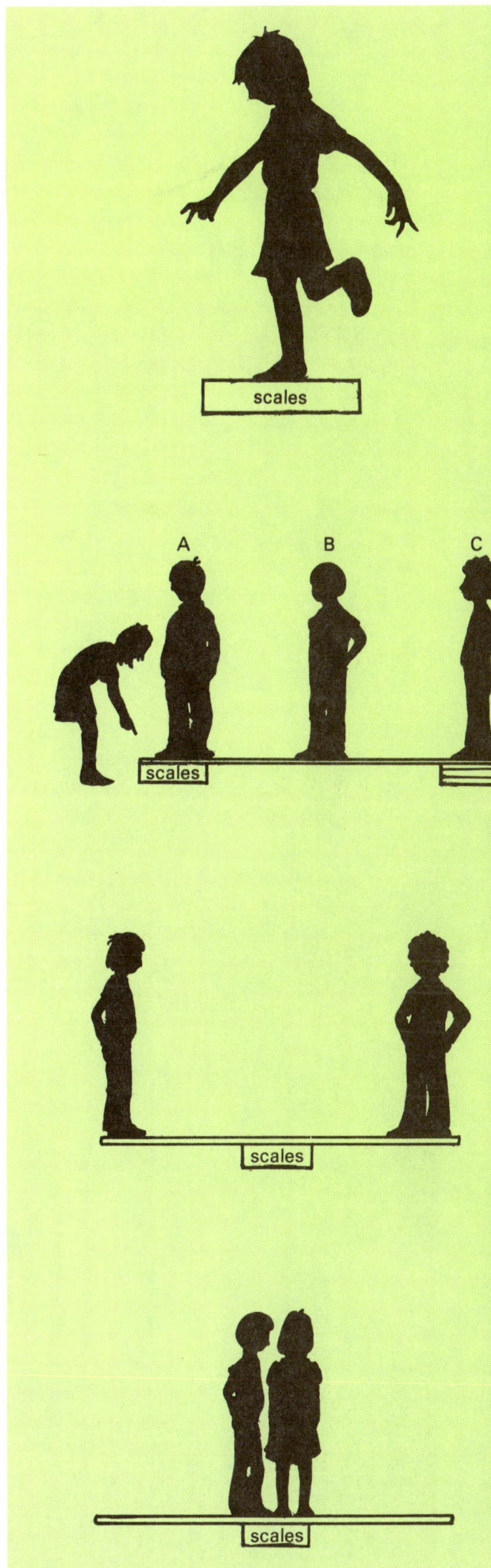

How Fast Does It Fall?

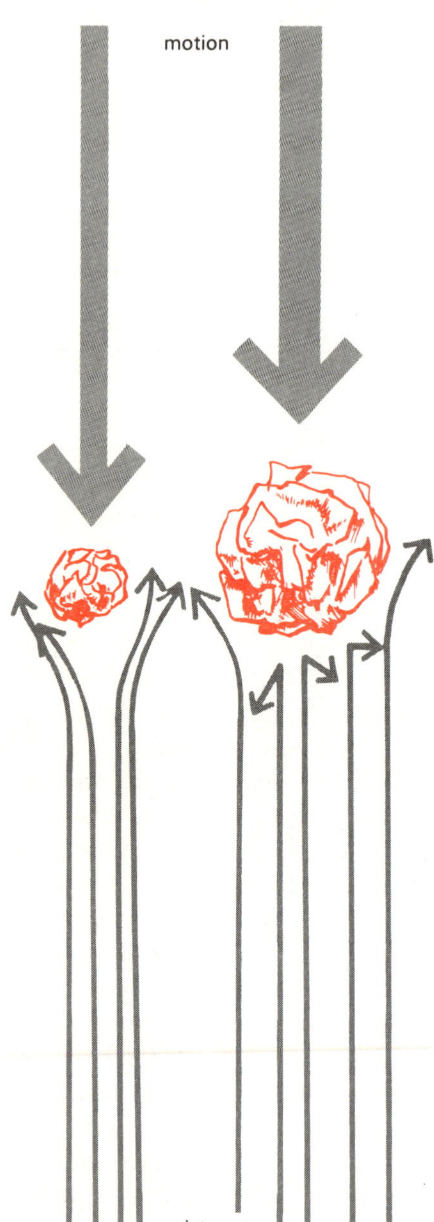

We have already talked about falling bodies and Galileo's famous experiment. Some 400 years ago in northern Italy, Galileo did an experiment from the top of a tower. He dropped iron balls of different sizes. They all hit the ground at the same time. And maybe you have done the same kind of experiment by standing on a chair and dropping two coins, like a dime and a half dollar. From all experiments like these we learn that the speed of a freely falling body is independent of its weight.

But just what does determine the speed of a falling body? There are two parts to the answer, so let's take them one at a time.

Galileo didn't stop with his experiment of dropping balls of different size. He asked questions about how fast things really do fall. And he discovered an important idea. If we apply a constant **force** or push on an object — and that's what the force of gravity does — then the object travels faster and faster, the longer it goes.

How fast something moves is its speed or **velocity.** We might measure speed in miles traveled in an hour (miles per hour) or in feet traveled in a second (feet per second). Now, the speed of a freely falling body is always increasing. After falling one second, an object is traveling 32 feet per second. After falling two seconds, it is traveling 64 feet per second. After falling three seconds, it is traveling 96 feet per second. And so on. An increase in speed is called **acceleration.** So we say that the force of gravity causes a constant acceleration in a freely falling body. At the end of each second of free fall, an object is traveling 32 feet per second faster than it did at the beginning of the second.

It would seem that a falling body would just keep traveling faster and faster till it hits the earth. But sometimes there is a second part to the answer, because air has some slowing effect on any object moving through it.

You can feel a wind blowing against you. And if the wind blows hard enough, you can even lean against it a little. If you build a big sail on a boat, you can make the wind push the boat along. If you fell out of an airplane, you would be moving through the air. The air would seem to be blowing up against you and trying to hold you back. And then, if you opened a parachute (I hope you would have one), it would give a bigger surface to push against the air and hold you back.

You have seen the effect of air resistance on the speed of falling bodies.

A feather doesn't seem to fall very fast. Maybe you have seen a snowfall when the flakes were big and fluffy. The snowflakes just slowly drift down. For comparison, a hailstorm is made out of just the same stuff — ice — but it is a nice round compact little ball. And when the hail hits the ground, it usually is traveling pretty fast.

You can do an experiment to show the effect of air resistance on a falling body. If you stand on a chair or stool you can drop two coins, like a dime and a half dollar, and see that they hit the floor at the same time. Now take a paper napkin and crumple it up into a very small ball (such as you are not supposed to throw in a schoolroom). If you drop the paper ball and dime together they will land almost at the same time. Now pull open the paper ball so that it is big and fluffy, and try again. Or you can compare the fall of two paper balls, one big and fluffy and one crumpled compactly together. You can see the effects of air on a falling body.

For a compact object — like an iron ball or a coin — air resistance does not have much effect until the speed gets very high. If experiments are done in a long glass tube with all the air pumped out (a **vacuum**), then there is no air resistance at all and a feather falls just as fast as an iron ball. When there is no air resistance, or when the effect of air resistance is small, we say that a falling body is a **freely** falling body.

You might like to consider that on the moon, objects must fall quite differently than they do on earth. The force of gravity on the moon is only a little more than one-sixth as great as it is on earth. So a falling body would speed up only about 5½ feet per second in each second of fall. That's a lot less than on earth. But the moon has no air around it to slow down falling bodies. So a parachute wouldn't be much good to you on the moon.

Try This!

It is easy to find things which will float on water, and other things which will sink to the bottom. You can make an egg do the very special trick of staying between top and bottom as if it were a submarine.

Do this experiment over the kitchen sink since you may spill some water. You need an egg, a nail, and a piece of cork. (If you can't find a cork, use a dry piece of wood like a toothpick.)

First try a glass of water not quite full. Put in it the nail and egg and cork. These behave just as expected. The egg and the nail sink to the bottom as in Figure 1.

Next try a glass of salt water. Add two level tablespoonfuls of salt to a glass almost full of water. Stir until the salt dissolves. Now add the nail and egg and cork. This time the egg should float with the cork, as in Figure 2. If it does not, dissolve some more salt in the water.

With a tall water glass you can get "real fancy" and combine both of these experiments into one. Fill a tall glass about half full with the salt water you used before. Now you want to pour a layer of plain water on top of the salt water without mixing them up. Unfold a paper napkin and put one corner of it down into the tall glass on top of the salt water. Hold the paper in place with your fingers and carefully pour in plain water until the glass is almost full. Carefully pull out the paper napkin.

Now use a spoon to let the egg down into the glass. The egg should sink to the bottom of the layer of plain water but float on the salt water — just as it did before. Tap it gently with a spoon to make sure that it really does want to stay there. If you add the nail and cork, you will see them behave as in Figure 3, and just as they did before.

Figure 1 Figure 2 Figure 3

How It Works

An egg is just a little heavier than an equal volume of water. It is more **dense.** It sinks. When we add salt, the salt water becomes more dense than the egg. The egg floats on salt water but sinks in plain water. Of course, you never can get salt water dense enough to float a nail.

Now I have a question for you. In your experiment you put a layer of plain water on top of a layer of salt water. Even if you wanted to, you could not reverse this and keep the layer of salt water on top. Why?

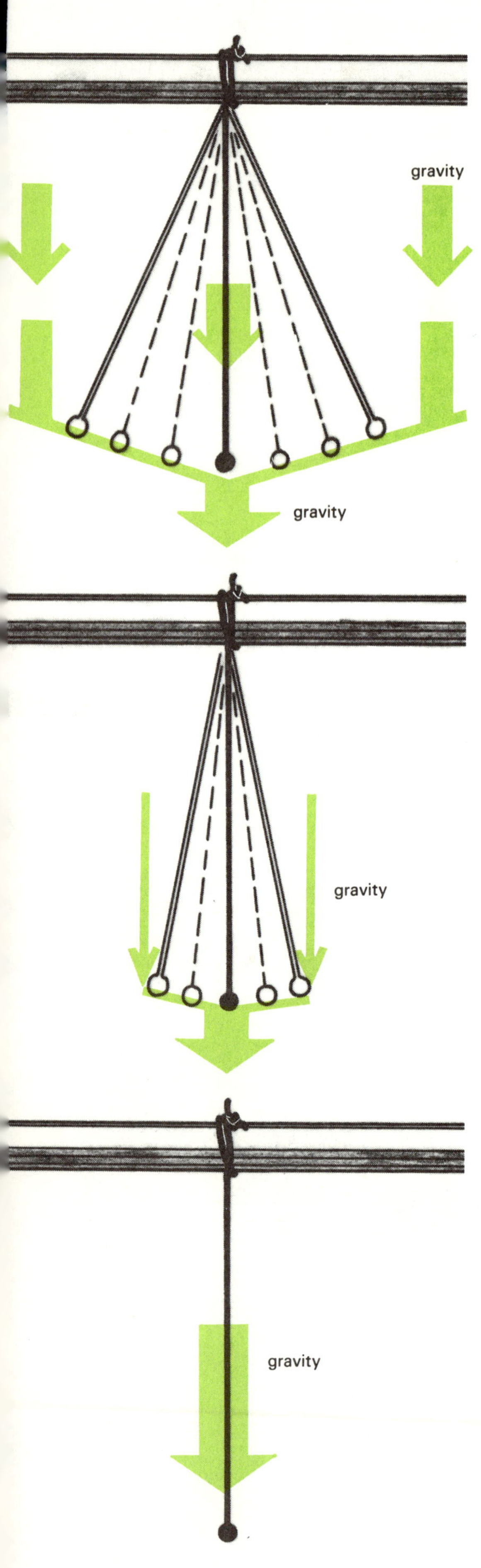

Watching a Pendulum

There are many things to wonder about in the world around us. Many of them are simple things in the pattern of nature which most of us do not even notice. Today I want to tell you about a man who watched a swaying lamp, and wondered, and discovered the idea of the **pendulum.**

We have talked before about the scientist Galileo who lived in Italy almost 400 years ago. When he was nineteen years old, Galileo was praying one day in a cathedral. As he finished his prayer and turned to leave, he noticed a lamp suspended by a long cord. The lamp had been lighted and left swinging to and fro. He noticed that the time of each swing seemed to be the same. He knew that his heartbeat was very regular, so he timed the lamp swings by counting his pulse.

It was a moment of discovery. Many people must have looked at swinging objects, but Galileo noticed something which no one before had realized. As long as the lamp remained swinging, even a little, the time for each swing remained the same.

Galileo hurried from the cathedral to borrow some string and a weight. He made what we today would call a pendulum. He tied one end of the string to a high place, maybe the limb of a tree, and tied the weight at the lower end of the string. When he pushed the weight, it would start swinging. As it gradually died down, the swings got smaller and smaller, just like the swings of the lamp. And just as he had noticed with the lamp, each big swing at the beginning took just the same time as a small swing when it had almost died down. But the time for each swing was much shorter than he observed for the lamp.

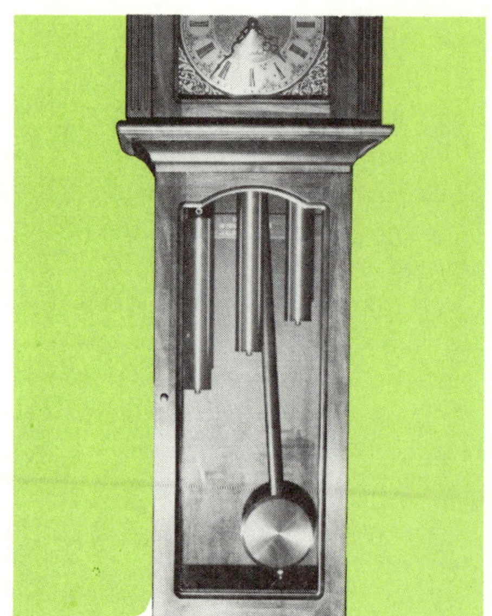

Why should it be so?

Galileo considered that there was a big difference between his pendulum and the lamp. The lamp was made of bronze. The weight of his pendulum was made of iron. So he tried a piece of bronze and then a piece of wood for his weight. He tied big weights and small weights onto his string. Always the time of the pendulum swing was just about the same.

What else about his pendulum was different from the cathedral lamp? Well, it was shorter. So Galileo lengthened the string of his pendulum. And then it all became clear. As he made the string longer, the pendulum swung slower and the time for each swing became longer. When he made the string shorter, then the time for each swing became shorter. Galileo had established the simple characteristics of the pendulum. The time for each swing, the **period** of the pendulum, depends on the length of the string or rope or whatever it is that holds the weight.

Maybe you can see why a pendulum works. A weight on a string wants to hang straight down. Why? Because of the force of gravity. When it is pushed off to one side, the force of gravity is still pulling straight down. But the string won't let go, so the weight has to coast down at an angle toward its bottom position. And when it gets to the bottom, its momentum carries it up the other side. It is really very much like coasting downhill on a bicycle and getting enough speed to climb part way up a hill on the other side.

Once you start thinking about pendulums, you can see them in many places. You may see a swinging lamp as Galileo did. You can find them in some clocks. (Why do you suppose they are sometimes used in clocks?) You can see them in playground swings. Any swinging object is a pendulum. And always the period of a pendulum depends upon its length.

Just think a little about how Galileo came to discover the idea of the pendulum. He watched a swinging lamp, he watched it carefully — and wondered.

Try This!

Let's build a pendulum. You can make one as Galileo did, and see how it works. You need a long piece of string or, even better, some strong thread. And you need a weight like a large iron nut or a small bolt. Almost any object will do for a weight except that it ought to be small and heavy. (The very best weight I have been able to find is a lead sinker out of my fishing-tackle box.) Tie your weight to one end of the string.

You can hold the string and let the weight dangle. If you push the weight to one side, it will swing to and fro, just like a playground swing. Let's call each swing from one side to the other a **beat** of the pendulum. How fast it beats depends on the length of the string. Does it beat faster with a longer string or a shorter string?

Let's do a careful experiment to find out just how the length of the string affects the beat of the pendulum. You need someone to help you. And you need a yardstick or a tape measure, and a watch with a second hand.

Tie the upper end of the string to something rigid and fairly high up, such as a doorframe. Outdoors, maybe you could use a low tree. First tie the upper end of the string so that the center of the weight is 20 inches below. Start your pendulum going with a push. While your helper looks at a watch, count how many beats the pendulum makes in a minute. It will make about 84 beats in a minute. And notice that it makes just about the same number of beats per minute when you start it as when it has almost died down.

Try different lengths of string, always measuring carefully from the upper fixed point to the center of your weight. Try 10 inches and 30 inches. Try other lengths. Try doubling the weight on the pendulum without changing the length. Does this have any effect on the beat?

Now you are on your own and I won't tell you how the experiment comes out. It will be more fun to find out yourself and even more fun to go on and invent different kinds of experiments.

And now here's a question you can answer from your experiments: How long should a pendulum be in order to beat 60 times a minute or once each second?

What Is Air Pressure?

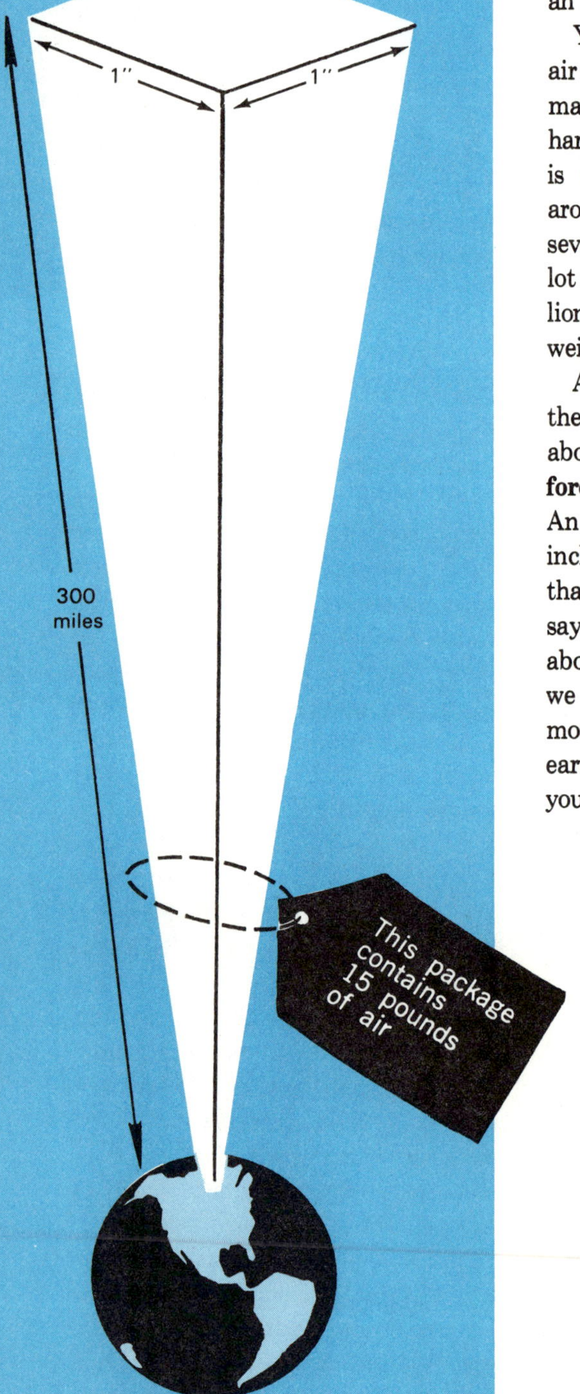

Question:
I have heard people say that you cannot build a pump that will lift water from a well more than 32 feet deep. Why is this?

It turns out, as you will see, that this is really a question about the nature of the air around us. You and I live on the surface of the earth but it is also true that we live at the bottom of an ocean of air.

You will remember that a quart of air weighs about 4/100 of an ounce. It may seem that this is so little as to be hardly worth talking about. But there is a lot of air in the **atmosphere** around and above us, and it extends several hundreds of miles up. That's a lot of quarts of air — billions and billions of quarts — so that its total weight is pretty great.

Actually, no one ever talks about the total weight of air. Instead, we talk about **pressure;** for example, the **force** or weight on each square inch. An easy way to see an area of 1 square inch is to draw a square with a ruler so that it is 1 inch on each side. When we say that the normal air pressure is about 15 pounds per square inch, what we mean is that all the air in the atmosphere above a 1-inch square on the earth's surface weighs 15 pounds. If you had a long, square-shaped plastic tube with 1-inch sides, standing on end and extending several hundred miles straight up, then all of the air in the tube would weigh about 15 pounds.

Also we have learned that air is a fluid like water. It can move and flow. So the air pressure does not just squash us down from the top. It also pushes against us on all sides and up from the bottom. That is why we don't feel that we are carrying a load on our shoulders.

There is an easy way to show the effect of air pressure. For this you need only an empty bottle and a sink full of water. Put the bottle on its side in the sink so that it is entirely filled with water (no air bubbles inside). Now turn the bottle upside down, carefully keeping its mouth below the water surface. What makes the water stay up in the bottle? It is the air pressure pushing down on the water in the sink so that it holds the water up in the bottle. Tip the bottle a little so that part of the mouth is above the water surface. Now a little air is pushed in directly. It rises to the top and pushes some water out.

Air pressure works for you every time you drink water — or milk or soda — through a straw. You say that you suck water up through the straw.

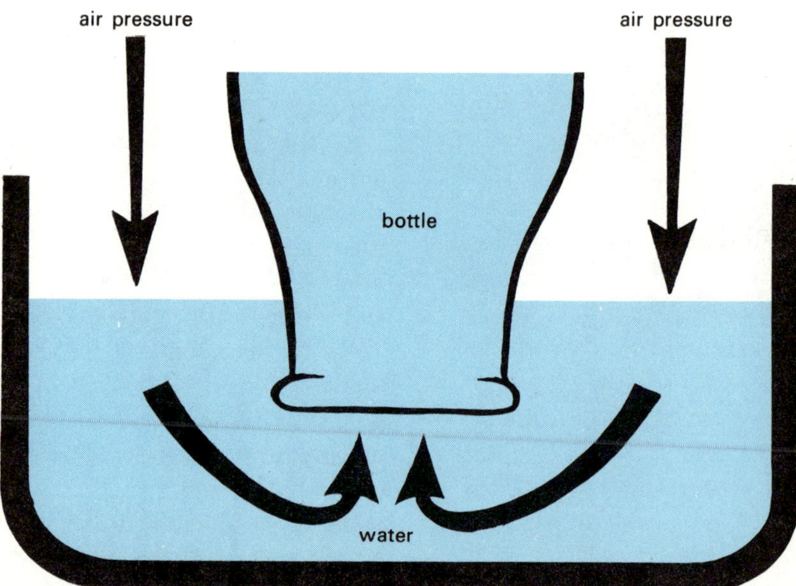

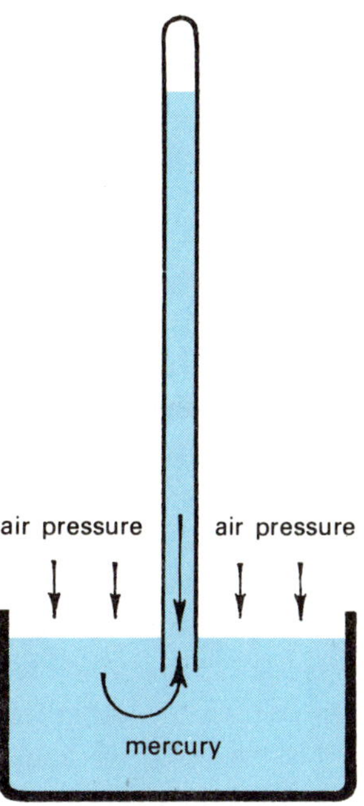

But what really happens is that you suck out some air or water at the top of the straw and the air pressure on the water in the glass pushes it up the straw. How long a straw could you drink through? Well, really, this depends upon how good a pump you can make out of your cheeks and mouth. But suppose you could build a very good pump. How high will it pull water in a pipe?

Way back over 300 years ago, scientists were puzzled by our question. No one had been able to build a pump which would lift water out of a well more than about 32 feet deep. Why should this be? Many people argued about it. Some said it was just a magic number of nature. One man, Evangelista Torricelli (tow-ree-chel-ee), got an idea and figured out how to do a simple experiment and found the answer. Torricelli knew that air had weight, and he got the idea that it was the pressure of air which pushed water up a pipe below a pump. If this idea was right, then it meant that air pressure was just enough to push water 32 feet high.

Now a 32-foot piece of pipe is pretty clumsy to handle, so Torricelli decided to use mercury instead of water. Mercury is a very heavy liquid which weighs about 13 times as much as an equal volume of water. So, if he was right, the air pressure should push mercury up about 1/13 as far as water — about 2½ feet or 30 inches.

Torricelli built a glass tube about 3 feet long and sealed at one end. He poured mercury into the tube and filled it all the way to the top. Then he put his finger over the open end of the tube and turned it upside down with the open end in a dish of mercury. When he removed his finger, the mercury level dropped from the sealed end but not all the way. And, what do you know, it stopped when the level reached a height just about 30 inches above the surface of the mercury in the dish. Torricelli showed that, even if the tube were longer or differently shaped, the answer was still the same. And if the tube were less than 30 inches long, the mercury stayed right at the top, just as the water stayed at the top in your experiment.

Torricelli explained his experiment just as we do today. If the tube is longer than 30 inches, then in the upper closed end there is nothing at all — or a **vacuum.** So the mercury column is not pushed down by anything except its own weight. But the air pressure is pushing down on the surface of the dish of mercury. So inside the tube the pressure of mercury pushing down is just exactly equal to the air pressure pushing mercury up.

So you see, no matter how good a pump you build, and even if you remove absolutely all the air from the top of a tube, you can't suck mercury up higher than 30 inches. That's as far as air pressure will push it. And for just the same reason, no one can build a pump that will suck water higher than about 32 feet.

Now you know that 32 feet of water is not a magic number. It is a way to measure the pressure of air around us.

Try This!

The Stack-of-Coins Trick

Make a neat stack of five or more pennies on a smooth table. Now, with finger and thumb, flick a dime along the table so that it hits the bottom penny of the stack. The bottom penny will fly out, leaving the rest of the pennies in a stack. You can repeat this, removing the bottom pennies one at a time.

How It Works

The moving dime hits only the bottom penny. Its momentum (energy of movement) is transferred to the bottom penny only. The rest of the pennies tend to stay right in place, and the stack just drops down. Notice that the dime bounces back a little. Notice that, if you push the bottom penny slowly with a dime, the whole stack slides along. The bottom penny must be hit sharply.

To Make a Trick

After you have practiced, you are ready to do a trick. Ask a friend to make a stack of pennies on the table. Then ask if you can have any penny which you can remove from the bottom without touching the stack. Then take a dime from your pocket and show your friend how you can trick pennies out of the stack.

Air Can Be Squeezed

Question:
What do people mean when they say that the tires on our car should have 30 pounds of air? How can a tire hold so much air?

You are right. The air in an automobile tire does not weigh 30 pounds. It weighs only a few ounces. But we really do say that tires carry 20 to 30 pounds of **pressure.**

In our last two chats we were thinking about several important ideas about air. First, it is something and has weight. Secondly, it is a fluid which flows even more easily than water. And from these two ideas there came a third, that air has a pressure pushing against us on all sides with a force of about 15 pounds on each square inch.

Air has another important property. It is **elastic** (springy or rubbery) and can be squeezed or **compressed.** You can easily show this by an experiment. You need an empty bottle. Wash off the mouth. Put your lips tightly over the mouth of the bottle. Blow into it as hard as you can, and then seal it by sticking your tongue into the mouth of the bottle. Now remove your tongue. You can feel some of the air in the bottle rush back out over your tongue. You were able to squeeze some extra air into the bottle so that it had a higher pressure than the air around you.

You can also do the opposite experiment. Suck on the bottle as hard as you can and stick your tongue in to seal it. You have removed some of the air from the bottle. The pressure inside is less than it is outside. And when you pull out your tongue you can feel some air rushing back in.

There are some interesting results of the elastic property of air. One is that the air around us is compressed by all the air above. And high up in the atmosphere the air gets thinner and thinner. At the top of Mount Everest, the world's highest mountain, there is only about 1/3 as much air in a quart and the air pressure is only about 5 pounds per square inch. A pilot uses this idea to tell how high a plane is. An airplane has an instrument called an altimeter. It has a dial which looks very much like a clock face, except that it reads in thousands of feet above sea level. What the altimeter really measures is the air pressure around the plane. The higher it goes, the lower the pressure becomes.

Beyond a few hundred miles out in space there is practically no air at all. So you see why a spaceship for people has to be very strong and tightly sealed to hold its air inside.

We can make use of the elastic property of air to make things springy — like automobile tires. Automobiles really ride on the air in their tires. At most service stations there are electric pumps which compress air into a tank so that it has a pressure of 75 or maybe 100 pounds per square inch. A hose is used to let air into tires. Most automobile tires are not pumped up to more than 30 pounds per square inch. (This really means 30 pounds per square inch more than the air pressure outside.)

So you see the answer to the question. Automobile tires may hold only a few ounces of air. A bicycle tire will hold even less. When we say that they should be pumped up to 30 pounds, we really mean a pressure of 30 pounds per square inch. Since most of us really are pretty lazy, we usually don't bother to say that "per square inch" part.

And though it takes only a few ounces of air to make a pressure of 30 pounds per square inch, the whole automobile rides on this elastic and springy air in its tires. If you don't believe it, just try riding in a car with a flat tire.

The Thermometer

Question:
How does a thermometer work?

Let's start by thinking about the problem of measuring temperature. We all know when something feels hot or feels cold. But how hot is it? Or how cold? Do you think that you and a friend could agree on just how hot or how cold a day it was today? And we might want to know how hot something is when it is too hot to touch. So we need a way to measure hotness or coldness, which is temperature.

In order to measure temperature it would be helpful to have a property (quality) of things which changes with temperature in some simple way. There are a number of properties we might use. By far the most useful is the property which many things have of expanding (getting bigger) on heating, and of contracting (getting smaller) on cooling. Some things do this more than others. As a result we can use these to make a thermometer to measure temperature.

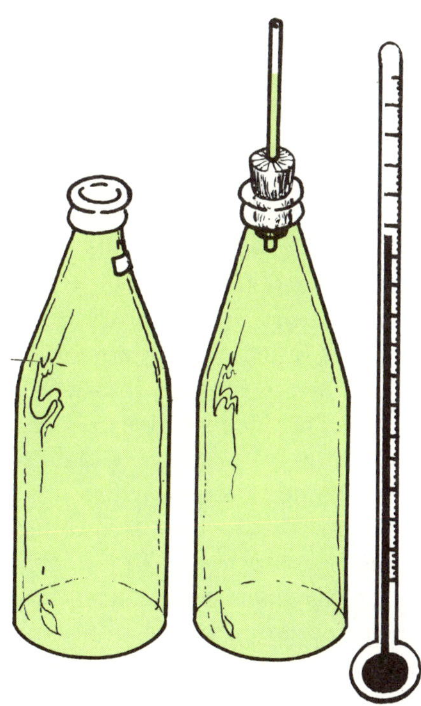

Let's make a simple thermometer. For this we will need a bottle with a narrow neck, like a catsup or soda pop bottle. We will start by filling it almost full from the hot water faucet, with water just too hot to hold your hand in. Let's carefully mark the water level with a piece of tape. And it might be a good idea to put a cap or cover, maybe just a piece of foil, over the mouth of the bottle.

Now we want to cool the bottle of water. Just letting it stand in the room will work. After a few hours, when the bottle no longer feels warm, look at the water level in the neck. It will have lowered a little. It will lower still farther if you put it in the refrigerator for a while. And it will come back up again if you bring it back out in the room to warm up. You have made a thermometer.

Why did your thermometer work? Well, the water contracted and decreased in volume by about 1/50th (one part out of 50) of its original volume. The glass bottle decreased a lot less, maybe about 1/200th of its volume. Your thermometer worked because water expands or contracts more than glass does, with temperature.

I'm sure you realize that your thermometer would not be very practical for everyday use. Let's think of some of the things wrong with it. First, it did not change very much with a big change in temperature. Second, it took a long time to change. And third, we couldn't use it to measure the temperature of an oven or a freezer. Why?

So let's think about how we could improve on the design of a bottle thermometer. If you try different kinds of bottles, you will find that the bottle with the narrowest neck works best. You will realize that if you make a small change in volume of water in the bottle, just by adding or taking out a teaspoonful, the level will change more in the bottle with the narrower neck.

"It's fifteen degrees freezingheit."

We might improve on the catsup bottle. Suppose we were to drill a hole through a cork and stick in a tight-fitting glass tube, as in the illustration. And suppose we pushed the cork in so that the water level rose up in the glass tube. Now if we warmed up the bottle, the increase in volume of water would force it a longer distance up the tube.

Now our bottle thermometer is beginning to look like a real thermometer.

There are still two other changes which would help. We could use another liquid, like mercury, which works better than water. And we could make the glass tube very, very small. Then we could use a smaller bottle. And that's how a real thermometer is made. It has a glass bulb, like the bottle, containing an expandable liquid and a much longer stem or glass tube that is very small inside. And it is made all in one piece of glass.

We don't know exactly who invented the thermometer. Likely it came from the ideas of many different people. And they probably went through the same reasoning that we did — except that when all this happened, about 300 years ago, they didn't have catsup bottles and soda pop bottles and refrigerators. And they didn't know as much about science as we do today.

Measuring Temperature

Question:
Why are there two different temperature scales like °F and °C? Which is correct?

This sounds like a simple and not very important question. But let me show you that it really is important.

Whenever we measure anything we must have some kind of units. In the United States and Canada we measure money in dollars. In France people measure their money in francs, and in Germany they measure it in marks. Their units not only have different names, they are also different in size from ours.

The important idea is that units of measurement are not magic. Most of them are just units of convenient size which someone thought up and which we are now used to. They are conventions, like driving on the right side of the road or having clocks with hands that always turn the same way. What units shall we use to measure temperature?

We don't know who invented the thermometer. We do know that it came into use in Italy about 300 years ago. But there was still a big problem. No two thermometers read the same. One might have read 60 when another read 70. So no one could tell anyone else just what the temperature was. It took about another fifty years for someone to solve the problem.

Back about 250 years ago in Holland, a man by the name of Gabriel Fahrenheit (fair-en-hite) made thermometers and other instruments. He improved the thermometer by using mercury as a liquid. Then he set about trying to find a way to make all his thermometers read alike.

He found that a mixture of salt and ice gave just about the lowest temperature he could get. And on any one thermometer, this mixture always gave the same temperature. It was already known that all healthy people had about the same temperature. Fahrenheit found that one of his thermometers in his mouth or his wife's mouth or his son's mouth read just about the same.

Now he had two temperatures he could count on. So he made a new batch of thermometers. He put them first in the mixture of salt and ice and made a scratch mark to show how far up the mercury went. Then, one by one, he put the thermometers in his mouth, or maybe in his wife's mouth, and made another scratch mark to show how high up the mercury went. The two marks on each thermometer were pretty far apart so he decided to divide up the distance in between. For some reason he decided to divide it up into 96 equal divisions. Each division he called a degree. In order to measure higher temperatures, he just added on more divisions of the same size.

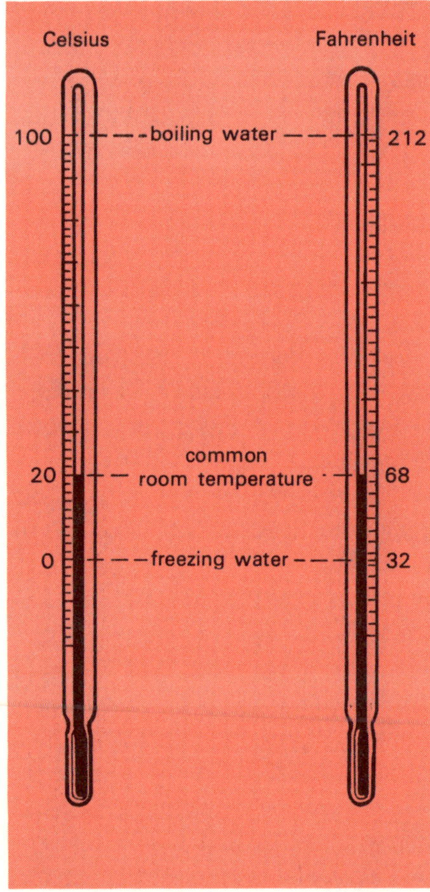

Later, Fahrenheit discovered that, on his scale, water always boiled at 212 and froze at 32. And it turned out to be easier and better to use these as check marks and divide the difference between them into 180 units or degrees (212 minus 32). This is the thermometer scale which some people still use today. It measures temperature in °F, meaning "degrees Fahrenheit."

Actually, Gabriel Fahrenheit made a mistake. On his scale, the body temperature of a healthy person was supposed to be 96. But that isn't quite right. Why don't you look up what a healthy person's temperature is supposed to be?

A different temperature scale was invented by the Swedish scientist, Anders Celsius. He thought it would be easier to divide up the difference between freezing water and boiling water into 100 units. This came to be called the Celsius or Centigrade scale. It has bigger units or degrees than the Fahrenheit scale, and it reads zero in freezing water and 100 in boiling water. The Celsius scale is used today in most of Europe, and by most scientists all over the world. It is written °C, meaning "degrees Celsius."

So you see why we have two different temperature scales. Which one is correct? Actually, both are correct. The choice is not a matter of rightness or wrongness. It is just a matter of which is easier. Most of the people in the world think the Celsius scale is much easier.

Water in the Air

Questions:
Why can I see my breath on a cold day but not on a warm day?
Why do my glasses get cloudy on a cold day when I walk into the house?
What does it mean when the weather report on the radio or television says that we have 60 percent relative humidity?

The answers to all these different questions depend on an important quality of the air around us. Air contains several different kinds of gases like oxygen and nitrogen in a mixture which is almost always the same, no matter where you go. And then there is another gas which varies a great deal. We call this water vapor, meaning that it is water in the form of gas.

Water vapor gets into the air because it is always evaporating from liquid water, from oceans and lakes and rivers, from any wet surface. You have seen that puddles of water on a street dry up when the sun comes out after a shower.

You can show evaporation by an experiment. Fill a drinking glass almost full of water. Carefully mark the water level by a piece of tape on the outside of the glass. Leave the glass where it will not be disturbed. (No fair drinking it!) In a day or two you should be able to see that some of the water has disappeared. It has evaporated and become water vapor in the air.

You might repeat the experiment and leave the glass in different places. You may be able to show that water evaporates more rapidly when it is warm or when the air is very dry or when air is continually moving across the water surface.

If you put a tight cap over the glass, or even a piece of cardboard, very little water will be lost. The air trapped in the glass above the water takes up all the moisture it can, then evaporation stops.

Now you know how water gets into the air by evaporation. But how does it get out? Again let's do an experiment. This one does not take so long. You will need two drinking glasses. (Aluminum drinking glasses or measuring cups would be best.) In one, put water which feels just a little warm to your hand. In the other, put some ice and a little water. Then carefully wipe off the outside of each glass with a towel and put each on a saucer. Leave them about ten minutes. Then look at the outside of the two glasses. One will be dry. The other will be damp on the outside and may have water dripping down into the saucer. Which one is this? And where did the water come from?

I can give you a hint that the experiment will work best if the air is not too dry. If you live in a desert, the experiment may not work at all.

Now we know how water vapor gets into the air, and how water can be taken out. There is one more important question. How much water vapor can the air hold? The answer to this depends upon the air temperature. Warm air can hold a lot more water vapor than cold air. The air behaves as if it were a sponge. And cooling air is like squeezing a sponge. The air next to a cold surface loses its water vapor which appears as water on the cold surface. When moist, warm air is cooled, its water vapor is squeezed out and changes to little droplets of water as in a fog. You cannot see water vapor (or steam, which is the same thing) but you can see the fog of little droplets of water which are squeezed out of the air by lowering its temperature.

At every temperature there is a different limit to how much water vapor the air can hold. We talk about water vapor in the air as humidity. At any one temperature, when the air has all the water vapor it can hold, we say that this is 100 per cent relative humidity. If the air has half as much water as it can hold, or is half "full," we say that this is 50 percent relative humidity (because 50 is half of 100).

The relative humidity changes from day to day. What do you suppose it is on a rainy, foggy day? And the relative humidity can change during the day. As the day warms up, the relative humidity is likely to go down. There is the same amount of water vapor in the air. But at higher temperatures, the air could hold more and so it is not so "full." People talk about relative humidity because it is a measure of how wet or how dry the air feels. In some places like New Orleans, Louisiana, the humidity is usually pretty high. In other places like Tucson, Arizona, it is usually pretty low. Why do you suppose this is?

I have answered one of the questions with which we started. Now can you answer the other two?

Try This!

Fun With Soda Pop Bottles

This trick should be done over the kitchen sink. You will need two empty soda bottles, some food coloring or ink, and some small pieces of paper.

First you should practice to see that you can balance a bottle upside down on another, as in the photograph.

The lower bottle must be sitting straight up, and both bottles must have smooth lips. If you can do this, you are ready to go.

Put about half a teaspoonful of food coloring (any color will do) into one bottle. Then fill it with warm water. Get the water as warm as you can without being hot enough to hurt.

Fill the second bottle with cold water. Hold a piece of paper over the top and turn the bottle upside down, as in drawing. Place its mouth over the mouth of the bottle containing the warm, colored water. Carefully pull out the paper and balance the bottles so they stand as in the photograph.

Look at what happens to the colored water from the warm bottle. Why does it rise into the upper bottle?

How It Works

Before we decide how the trick works, let's do it in a different way. Repeat it as you did before, but put the coloring in cold water in the lower bottle, and put warm water in the upper bottle. This time nothing happens. Holding the mouths of the bottles together with both hands, turn them upside down. Now the bottle of cold, colored water is on top. What happens?

Really, you have done three experiments. In each of them, the coloring was just to show you what the water was doing. Cold water is more dense — a bottle of it weighs a little more than a bottle of warm water. So cold water tends to sink, and warm water tends to rise. Both things happened in your first experiment. There was warm water rising and cold water sinking. As the warm water rose, it got all mixed up to give irregular currents and the pretty patterns of color in the upper bottle.

Starting and Stopping

Question:
I had to stand up on a bus the other day, and this is what I noticed. When the bus started, I had to hang onto a seat to keep from falling toward the back. But when the bus stopped, I had to hang on to keep from falling toward the front. Why is this?

Your question has to do with what is called **inertia.** When something is at rest, it wants to stay at rest. And when something is in motion, it wants to stay in motion — and in a straight line, too. An outside force has to act to start something moving, or to stop it once it gets in motion.

When the bus started, your feet stuck to the floor. There was a force outside of you acting to move your feet forward with the bus. But there wasn't any force (except a little one due to the air in the bus) acting on the upper part of your body. If you had not held on, your feet would have gone forward with the bus and you would have fallen backward in the bus. When the bus got up to a constant speed, you were moving just as fast as the bus. Then the driver put on the brakes, the bus slowed down, and your feet slowed down. But the rest of you kept right on going forward, so you had to hang on again.

What do you suppose would have happened if you had been on roller skates?

There are many experiences like this which depend upon inertia. Maybe you have noticed the funny feeling in your stomach when on a fast roller coaster going over a little hill, or in a fast elevator which suddenly started downward. Your stomach is not held in place very tightly inside of you. When an elevator starts downward, your stomach wants to stay right where it is. That means that it pushes upward and gives you a funny feeling inside.

Here is a little trick which depends upon inertia and is fun to learn. You will need a card and a coin. A playing card and a nickel work very well. Lay the card on the tip of one finger of your left hand and put the coin on top so it is balanced. Now give the card a snap with the middle finger of your right hand, as in the illustration. If you will practice a little to get just the right snap, the card will go flying away and the coin will stay right there on your finger. It won't work very well unless the card is smooth and not wrinkled.

Can you explain why the trick works?

Try This!

See Through Your Hand!

Roll a sheet of writing paper into a tube. Hold the tube up to your right eye with your right hand and look through it. Then hold your left hand against the tube as shown. Keep both eyes open. Look through the tube at something far away. See! You're looking right through a hole in your hand!

How It Works

When your brain receives messages from your eyes, it must combine the two pictures which your two eyes see, into one picture. If it didn't, you would see two of everything which would be very confusing, and make counting twice as hard. To help you out, your brain overlaps the pictures before it shows them to you.

When you look through the tube with one eye and at your hand with the other, your brain puts the two pictures together so you appear to be looking through your hand.

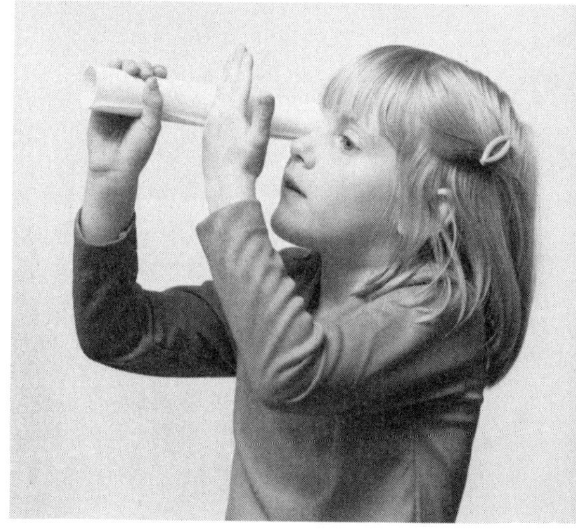

Machines Are Energy Changers

Question:
When I look at most things that work, I can see that it takes something to make them run. An automobile won't run without gasoline. An electric fan won't run unless it is plugged into an electric outlet. And an electric light bulb won't give light unless it is screwed into an electric socket. I can't think of anything that runs all by itself. Is there anything that does?

From your observations about an automobile and an electric fan and a light bulb, I think you realize a basic idea of science. You can't get work out of a machine without spending something to make it go. What you need to spend is called **energy.** Now, energy isn't stuff that you can handle or weigh. And it appears in different forms. So first let's make sure we understand about energy by thinking about some different kinds of machines.

Any object that moves has **mechanical** energy. The turning blades of an electric fan have mechanical energy. Because they are angled, they push the air and give it mechanical energy. What makes the fan blades turn? You have already observed that the electric motor of the fan has to be plugged into the electric line. It is run by electricity or **electrical** energy. An electric fan is a machine. We put in electrical energy and get out the mechanical energy of moving air.

You also observed that an electric light bulb needs electrical energy to make light. Light is another form of energy. We call it **radiant** energy. And so, even though it has no moving parts, a light bulb is a kind of machine. It converts electrical energy into radiant energy.

Where do we get electrical energy? Well, most of us get it through power lines with wires coming into our homes. And somewhere these electric lines connect to a powerhouse. You probably have one not too far away from where you live. Most of them would be glad to arrange a visit for you or your class. You can get to see how electricity is made.

First there is a generator which really is just like a big electric motor — except that it works in reverse. By making the generator turn, we can generate electricity. To turn the generator we need mechanical energy. Now, different power plants use different means of getting mechanical energy but most of them depend on steam. If we heat water to make it boil it expands over a thousand times to make water vapor or steam. And we can make the rushing steam drive a turbine which is really just a fancy kind of fan with many blades. The turbine will drive the generator. To make all this work, we need heat to boil the water, and **heat** is another form of energy.

How do we get heat? Again there are many ways, but most commonly we get it by burning something, a fuel like coal or oil or gas. Burning is a chemical reaction. Really there is no loss of weight of stuff in burning, even though the material is very much changed and most of it disappears into the air. But energy is released. So we say that a fuel has stored-up **chemical** energy. Energy can be stored in various ways. For example, a stretched rubber band has stored-up mechanical energy. But the very best way to store energy is in the form of chemical energy. Some things like flashlight batteries or gunpowder or gasoline have lots of stored-up chemical energy.

We have talked about the five forms of energy: **mechanical, electrical, radiant, chemical,** and **heat.** We can build machines that will convert any one form into any other form of energy. It is just as if we had five different kinds of money, say dollars, quarters, dimes, nickels, and pennies. The idea is that the different kinds of energy are a little like different kinds of money — they can be changed from one kind to another.

Now, there is also a second idea that always goes along with the first. We can change energy from one form to another. But whenever we do, and whether we like it or not, some of the energy is transferred into heat. And heat is a special kind of energy. Once we get our energy as heat we never can get it all changed back into any other form. Try touching (carefully) an electric motor, or a light bulb, or an automobile engine when they are working. They are machines which are not trying to make heat, but when they are working, they always do. If we wanted to, we could make some of this heat do work, too. But heat just sort of gets frittered away and we never can get all of it to do useful work.

So changing forms of energy isn't as simple as changing money. Or maybe you can imagine what would happen if changing different kinds of money worked like changes in energy. Suppose that pennies were like heat and no one wanted pennies. Suppose whenever you got change for a dollar you would always get some pennies. And even if you saved your pennies very carefully, a bank would give you only one dollar for 150 of them. It would make it even tougher to live on your allowance, wouldn't it? I know this sounds pretty silly in terms of money — but that's just the way it is with energy.

There are very few times in science when we say, "It can't be done." But there are some things we know we can't do. Here's one. Why can't we build a generator to make electricity, let the electricity drive an electric motor, and let the motor drive the generator? Of course nothing would get it started, but suppose that we started it with another motor. Why wouldn't it keep running? Because both the motor and generator would produce a little heat (even if we added a steam engine to make use of the heat, some of the heat would be lost) and our motor and generator would slow down and finally stop.

What we have just talked about, if it worked, would be called a **perpetual motion** machine — a machine that would run all by itself forever. The United States Patent Office used to accept patents on ideas for perpetual motion machines. But it doesn't any more. This is one thing in science that can't be done.

Try This!

You can perform a scientific experiment to find out if salt makes water harder or easier to freeze.

Take two large tin cans — one-pound coffee cans will do very well. Fill each can to the same level, about half full of water. Then put about two-thirds of a cup of table salt in one can. Stir until the salt dissolves.

If it is freezing outside, put the two cans out on the porch overnight. Or put both cans in your freezer, or in the freezing compartment of your refrigerator.

After several hours, examine the cans. (Or maybe you will have to wait overnight.) What do you find?

In this experiment you have kept all the important things about the two cans the same **except** that one has salt in the water. The cans are the same. There is the same amount of water in each one. Both were kept at the same temperature for the same length of time.

So if you find any difference in the cans after several hours of freezing, you can say, as a true scientist: "Ah! When you add salt to water, this is what happens!"

Add 2/3 cup table salt to the water in this can.

Stir salt until dissolved.

Let both cans sit for several hours or overnight at freezing temperature.

Something New Under the Sun

By Rebecca Roman
and Jack Myers

That big old ball of fire, the sun, is doing a fine thing for us, just making sunlight. Let's consider some of the important things sunlight does for us and how we are trying to learn to use it better than we do.

We usually think of sunlight as light which allows us to see. It also keeps the world warm enough for us to live in. Light is a form of energy, like heat or electricity. And when light is absorbed, it usually is turned into heat.

Scientists often talk of sunlight as being solar energy (energy from the sun). We should be able to learn how to make this solar energy do more useful work for us.

The rate at which we get or spend energy — or do work — is called power. How fast do we get energy from the sun? A useful way to answer the question is to say that at noon, when the sun is right overhead, we get just about one horsepower on each square yard. Actually, there are always big losses in changing energy from one form to another. So the practical amount of usable power we can get from the sunlight falling on one square yard is only a small fraction of one horsepower. This means that you would have to catch almost all the sunlight that falls on the roof of your house to get enough power to drive a small automobile. And how would you drive it at night?

So now we know the basic difficulties of using sunlight: it is too "spread out," and we get it only about half the time. Then why do we want to put it to use? The answer is that every year our demand for power gets greater. But the world has only a limited supply of coal and oil, our power fuels. So we must find new sources of power.

We have one very successful kind of machinery for using sunlight. This is the green machinery of plants. It works very quietly to make, out of water from the soil and carbon dioxide from the air, materials from which a plant is made. When the plant material is dried, as in hay or straw or wood, we can burn it and produce heat energy. Plant material has stored-up chemical energy which can be released — it is a fuel.

For almost two hundred years scientists have been trying to find out just how the green plant machinery works. Perhaps someday we shall learn how it works, and how to do the job still better, but we have a very long way to go. When a farmer grows a field of corn, less that 2% of all the sunlight which fell on the cornfield is stored up.

A second way in which we might use solar energy would be to turn it into heat. Houses have been heated by sunlight even in northern climates. Imagine that the roof of your house faces toward the south and has a black metal roof covering. Black is chosen because it absorbs the most light and will get warm when sunlight falls upon it. Normally this warmth would be lost in winter to the cold outside air and by infrared or heat ra tion, which is like light except that our eyes cannot see it.

But suppose that we cover the roof tightly with large panes of window glass held a little way above the black metal. Now we have trapped the heat of the black metal underneath. Light can get in through the glass but the infrared rays can't get back out through the glass, and the outside cold air can't get in. The air between the glass and black metal gets hot, and all we have to do is to continually blow it into the house. We could even store the heat by filling the basement with stones and blowing some of the hot air over them during the day. Then at night we could blow air from the house over the stones and back again, to keep the house warm.

The design of the roof we have imagined, the large black plate and glass cover, is called a flat plate collector. Sometimes water, instead of air, is pumped through a flat plate collector. Many homes are using this idea for solar hot-water heaters.

In turning sunlight into heat, there is another idea we can use. We can concentrate the sunlight. Suppose we wanted to boil water in a black kettle which we might hang from a

solar-heated office building

clothesline on a sunny day. Just hanging there in the sunlight, the water would never boil. There is just not enough power. But suppose you were to stand behind and a little to one side of the kettle, and hold a flat mirror in position so that it would bounce sunlight back onto the kettle. That would almost double the power on the kettle. And suppose I were to stand just on the other side doing the same thing with another mirror. That would almost triple the power. If we can get enough mirrors, we can really heat up that kettle.

One idea like this is being tried in building what is called a power tower. Many large mirrors are arranged in rows to reflect sunlight onto a black boiler. In order to have room for a large number of mirrors the boiler is held up on a tower, maybe 500 feet high. That way sunlight can be concentrated about 1,000 times. The idea is that water flowing through the boiler will be turned into steam. The steam will drive a turbine. And the turbine will drive a generator to make electricity.

The power tower is based on the good idea of concentrating light with mirrors. However, building a big one is not so easy. It is just hard to imagine a big field, several hundred times larger than a football field, filled with thousands of large mirrors. Each mirror must be held up on a post. And each mirror must have a motor to turn it so that it properly follows the sun from morning to night. You can see that the power tower has a lot of problems to be solved.

Still another way to use solar energy is to convert it directly into electricity. In some kinds of material, like silicon, light drives out electrons and makes electricity. Crystalline silicon is cut into very thin plates, each a few inches square. These have come to be called photocells or solar cells or solar batteries. We know that they work very well because they have been used in big panels on our satellites to power motors and radio transmitters. The big problem is that solar cells have been so hard to make and so expensive. Naturally, people are trying to build them more cheaply. If that problem can be solved, solar cells will give us a neat and quiet way to make electricity from sunlight.

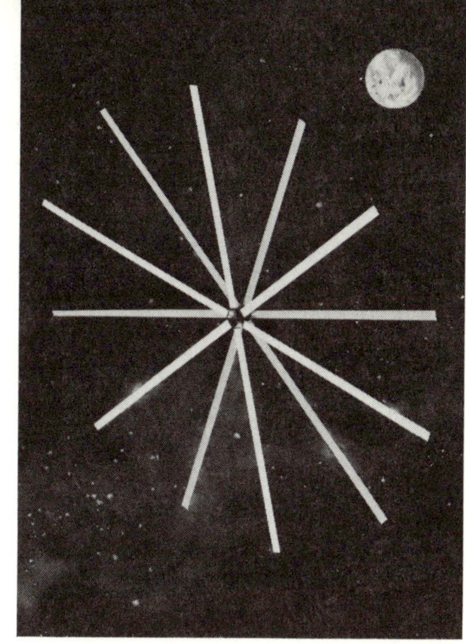

Solar energy may power many spacecraft.

For a very long time plants have used sunlight to do their chemistry. Today we have learned to make solar cookers and solar furnaces and solar batteries. What can we learn to do tomorrow? It will be fun to see.

There are still other ways to use solar energy. We can make windmills to use the energy of winds which are made by solar energy in heating air. Maybe we can even make some kind of machine to use the energy in ocean waves which are made by the winds.

You can see that there are many ideas about how to use solar energy. All of the ideas raise new problems in science and engineering which need to be solved. I hope we are able to solve all of them. We need to use solar energy in every possible way.

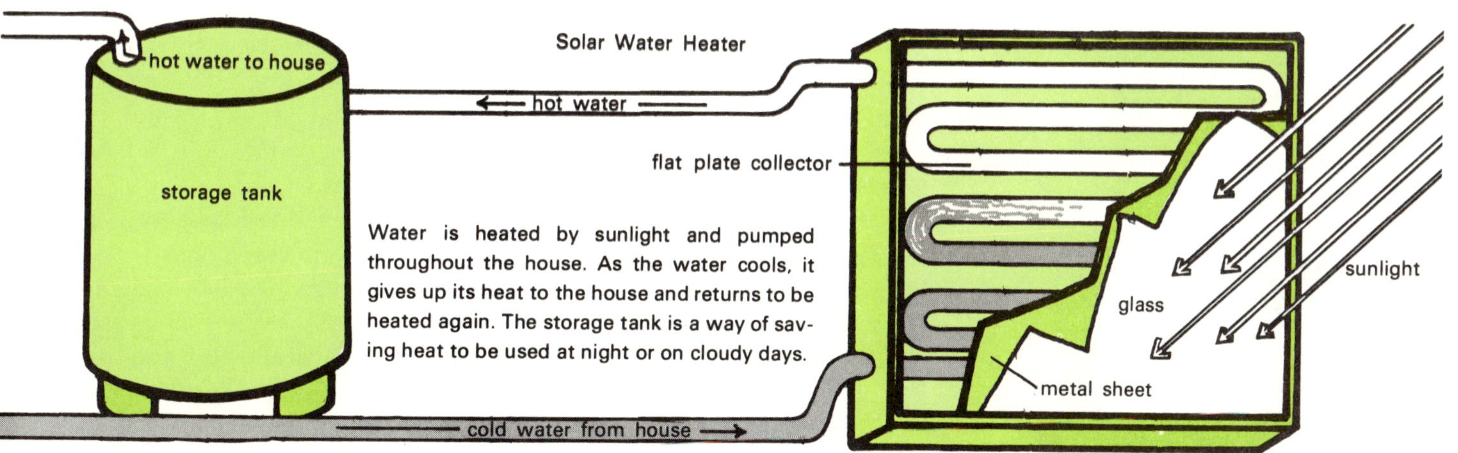

Solar Water Heater

Water is heated by sunlight and pumped throughout the house. As the water cools, it gives up its heat to the house and returns to be heated again. The storage tank is a way of saving heat to be used at night or on cloudy days.

Making a Compass

By Glenn O. Blough
Professor Emeritus
University of Maryland

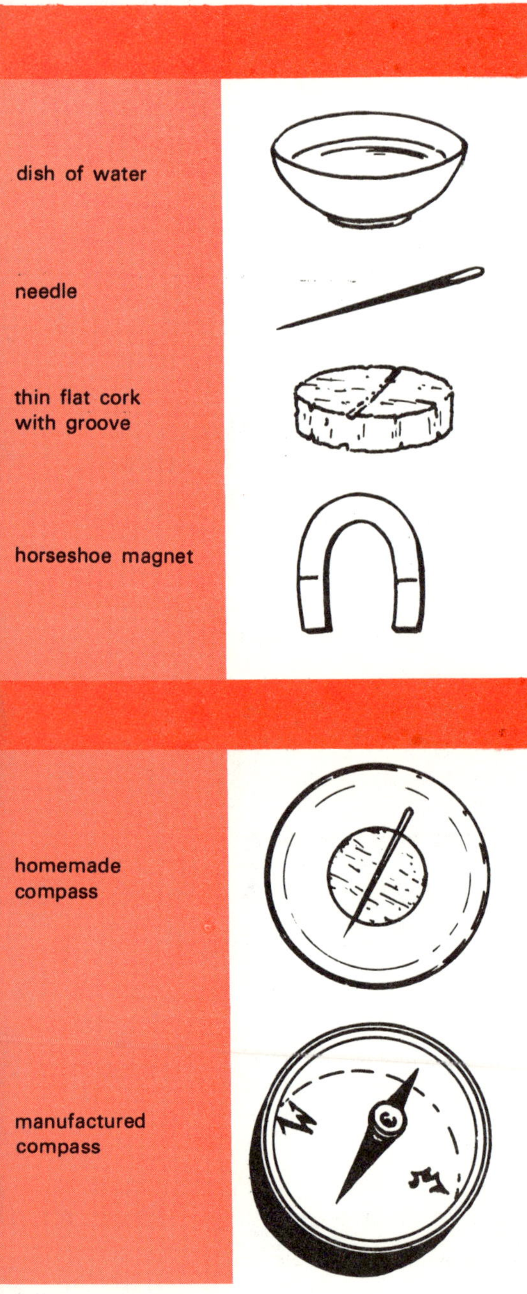

Every now and then people get mixed up in their directions. If you asked a dozen people which direction is north, they might all point in different directions. Right now as you are reading this, can you point to the north? Are you sure you are right?

It has always been important to know directions. You can't travel long distances by air or sea without being sure of them. Hunters, trappers, explorers, and others must be sure of their directions. The magnetic compass is an instrument that has been used for thousands of years by people all over the world to help find their way.

Suppose you point to the direction that you think is north. Now you can make sure you are right if you know how to use the things that are pictured here.

You probably have seen small magnets that are shaped like the horseshoe magnet in the picture. Magnets have other shapes, too. Sometimes they are shaped like a U and sometimes like a bar. The bar magnet is like the horseshoe and U-magnet except that it is straight. All magnets are very much alike, no matter what shape they are. They all pull things made of iron and steel toward them. They all have two ends or poles. One is called a north pole and the other a south pole. Some can pick up heavier things than others. That is because some have a stronger force than others. This force is called **magnetism.** This magnetism is strongest at the poles of a magnet, and weaker in the part between the poles.

If you have a magnet, you can make another magnet from it. For example, if you have a magnet you can make another magnet out of an iron pin. (Some pins are brass and won't work.) You can do this by rubbing the pin on the magnet several times. Then you can test the pin to see if it is a magnet. You can do this by trying to pick up another iron pin with the one you have been rubbing on the magnet. Why don't you try this?

There is a very interesting thing about a straight magnet like a bar magnet. If you hang it up by a thread so that it can swing around easily, it will stop swinging around after a while and point north and south. A steel needle will do that, too, if it is made into a magnet.

First make the needle (a big one about two inches long is best) into a magnet by rubbing it on one end of the bar magnet. Don't rub it back and forth or sideways. Rub the needle in one direction only from the center to the end of the magnet, holding it parallel to the magnet. Rub it twenty or more times. Now try to pick up an iron pin with the needle. If it picks it up, you know that you have made the needle into a magnet.

You can make a compass if you can support the needle so that it swings freely. Try to find some way to float the needle on a dish of water. You can lay the needle on a flat piece of cork. Or you can attach it to a dry twig of wood or to a little piece of plastic foam. Once you have your needle magnet floating freely on a dish of water you have made a **compass.**

It will be easy to see that your needle will move and point toward a magnet. Now take away the magnet and anything made of iron. If your compass can swing freely, it will always end up pointing north and south.

Other needle or straight magnets will behave in just the same way. If they can swing freely they will end up pointing north and south. The earth itself is a giant magnet lined up to have a north magnetic pole and a south magnetic pole. Its magnetism is strong enough to make a compass needle line up in the same direction.

If you have a globe that is round like a ball, you may find a north and south pole marked on it. These poles are the **geographic** North and South Poles. The **magnetic** poles of the earth are near these Poles.

But how can you tell which end of your compass needle is pointing to the north magnetic, and which pointing to the south magnetic pole? There is more than one way to discover this. There is an easy way if you have a regular pocket compass like the one pictured. It works just like your homemade compass except that its needle is supported between two pivots. The end of its needle which points north is usually dark-colored. So the end of your compass pointing in that same direction is also pointing north. Of course the other end is pointing south. If you face toward the north, then east will be toward your right and west will be toward your left.

There are other ways besides the compass to tell which way is north. Can you think of some?

It Seems Like Magic

By Ali R. Amir-Moez
Department of Mathematics
Texas Tech University

Here is an interesting experiment with triangles — three-sided figures. Three are shown.

Choose one shape. Cut a figure like it from paper. Make it larger than the figure shown.

Mark the corners, A, B, and C, as in Figure 1. The paper triangle is to be cut in three pieces as shown by the dotted lines. The three letters will be on the original corners.

No matter what size or shape triangle we choose, we see that the three corners (angles), A, B, C, fit together on a straight line as in Figure 2. The corners A, B, and C can be put together in any order, and this is true.

Experiment with other triangle shapes. Two others are shown.

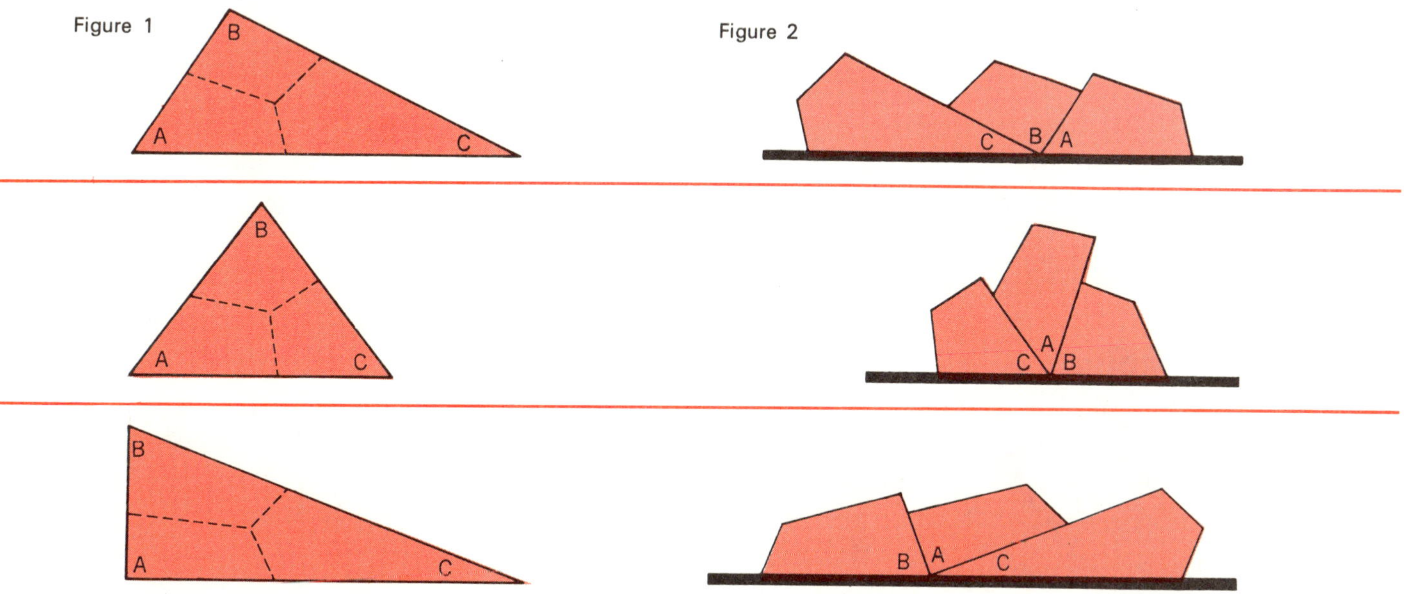

Electricity

By Glenn O. Blough
Professor Emeritus
University of Maryland

You have never seen electricity. Neither has anyone else. But you have seen it at work.

Electricity makes electric motors run and this is very important. There is an electric motor in a vacuum cleaner, in an electric fan, in a washing machine, and in many other things in your home. Electric motors are used in factories and many other places.

Electricity can make light. It lights our houses and schools and other buildings. It lights our streets and signs.

Electricity can make heat. Sometimes we use it to iron clothes, sometimes to toast bread, sometimes to make coffee and waffles. And sometimes we use it to heat our houses.

Electricity helps us to send messages by telephone and telegraph. It runs radios and televisions.

Even though we can't see electricity, we can learn about it in a simple experiment. We need the several different things connected up as you see them below. Not all of these are easy to get. However, we can easily do the experiment just by thinking about it.

As a source of electricity we will use a common kind of battery also called a dry cell. There are different kinds of batteries but in one way they are all alike: they contain chemicals which can react together to make electricity.

If you have used a flashlight you know that it contains small dry cell batteries. They give the electricity that makes the flashlight bulb give light. Our diagram shows the same parts needed in a flashlight but the parts are larger and spread out so we can see how they work.

In your automobile, the electricity comes from a storage battery. The storage battery gives electricity to start your car, to light the lights, to work the windshield wipers, and for other things.

Electricity travels over wires to your house. It may come from a dam or from some other place where there is a dynamo. Dry cell batteries, storage batteries, and dynamos are all sources of electricity.

Look at the picture and see what else you can discover about electricity. There are two places at the top of the battery. They hold the copper wires. The cover of cloth or plastic or rubber has been taken off from the ends of the wire where it is fastened to the battery. This is very important. Electricity will not go through this cover.

Now look at the light bulb and its socket. Do you see that the socket also has two places to hold wires? Bare wires are fastened to each of these places. This will let electricity travel into one side of the light and out the other. When it goes into the light bulb, it travels through a very, very thin wire. This makes the tiny wire so hot that it glows and gives off light.

Can you see where the wire goes from the light? It goes to the switch. The switch also has two places to hold the wires. Bare ends of wires are fastened to each of these places. One of the wires goes back to the battery.

Let's see if we can follow the path that electricity can follow along the wire. It comes from the battery and goes into the wire. It travels along the wire and goes into one side of the lamp. It goes out through the copper wire and goes into one side of the switch. Here it stops. Can you tell why?

The switch is like a drawbridge. When it is closed, the electricity can travel through. When it is open, as it is in the picture, the electricity stops. The electricity must have a complete path or it will not move at all. A more common way to say this is that electricity must have a complete circuit.

As soon as you close the switch, electricity goes on its way through the switch, back to the other side of the dry cell, and the electric light goes on. If you think about this you know the following things about electricity:

(1) It needs a complete path or circuit, or it will not travel.

(2) It can go through copper but it does not go through cloth or plastic or rubber.

(3) A battery can make an electric current.

(4) When electricity makes tiny wires hot enough, they give off light.

You can learn more about electricity by looking around the house. But now you must be very careful not to touch the wires. Your dry cell puts out electricity only at a low voltage and can't hurt you. But the electricity in your house works at a much higher voltage and can give you a bad shock. So no one ever touches a bare wire that is connected up.

See if you can find where the wires that carry electricity come into your house. Look near this place and you will find the electric meter. It measures the electricity that comes into your house, and tells you how much you use.

Look carefully and you will find the wires that carry the electricity. Are there two of them? Look at an extension cord that is not plugged in. It is usually easy to see that there are two wires covered with cloth or rubber or plastic. Look at the two prongs at the end of the wires. These fit into the socket in the wall. Two wires — two prongs in the socket — they are needed to make a complete circuit for the electricity.

The wires are made of copper because copper is a good carrier of electricity. They are covered with cloth or rubber or plastic. These are not good carriers. A complete circuit needs copper or some other good carrier of electricity (we call this a good conductor) and it needs something that is not a good conductor (we call this an insulator) to keep the electricity from going where we don't want it to go.

Usually, electricity in your home is something you just take for granted. But once in awhile a power line breaks in a storm, and your whole neighborhood is dark. Then you realize how very useful electricity is and how much we depend upon it.

Strange but True

By Ali R. Amir-Moez
Department of Mathematics
Texas Tech University

Let's play with a triangle. I guess you know that a triangle is a three-sided flat figure. Cut a triangle of any size or shape out of a piece of paper, Figure 1. Be sure to cut it carefully. Find the middle of each side by putting the two ends of each side together carefully, and folding that side. The crease is exactly the middle of that side, Figure 2. Now use your pencil and mark each crease.

If you draw lines connecting the middles of the three sides, you will get four triangles. Let us number them, 1, 2, 3, and 4, Figure 3. These triangles are of the same size. We test it by cutting them from one another, Figure 4. Now we see that 1, 2, and 3 fit one over another precisely, but you have to twist the triangle 4 upside down to make it fit over the others, Figure 5. Try this a few times with different size and shape triangles. It works every time. It is strange but true.

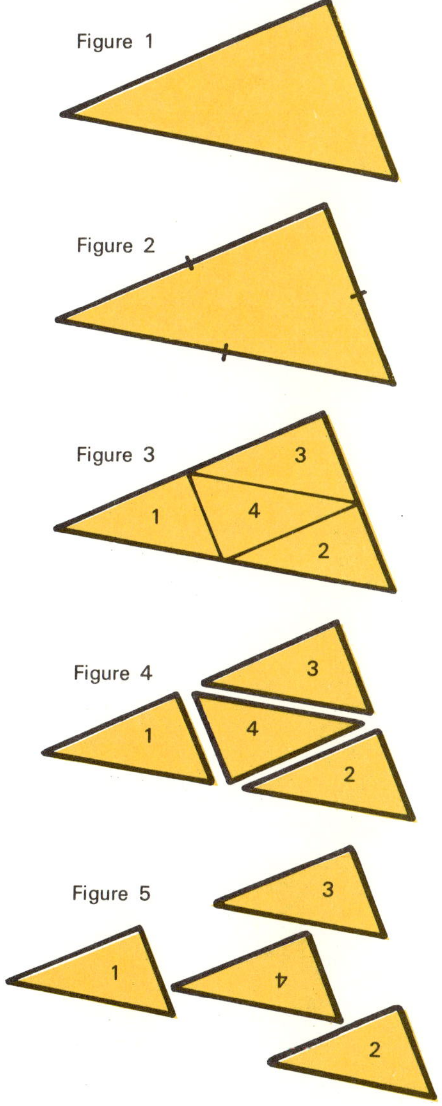

Look at the Machines!

By Glenn O. Blough
Professor Emeritus
University of Maryland

Machines push, they pull, they lift, they cut, and they do all sorts of other things. Look around you, and you will see them working to push dirt into ditches to fill them up. You can watch them pull heavy loads and lift automobiles and cut and carry and bend and dig.

What makes these machines go? Muscle force makes some of them move. It makes some lawn mowers move to cut the grass and bicycles move to carry people and steering wheels move to guide automobiles. But most of the machines you see every day are moved by a stronger force than muscles.

Electricity makes some of them move — washing machines, elevators, and electric fans. When you find a machine with an electric motor, you will know that it is electricity that moves it.

Steam makes some of them move. Steam runs steam turbines to make electricity and steam locomotives to pull tons and tons of freight.

An explosion makes some of them move. An automobile motor uses explosions of gasoline to drive its pistons and make it go.

Wind moves windmills to pump water and moves sailboats. If you look around, you will discover other forces that make machines move.

Now let's look at the machines. Some are very complicated. Some are very simple. The complicated ones like power saws and electric mixers and all the others are made up of a number of simple machines. What are they? They are levers, pulleys, wheels, inclined planes, screws, and wedges. They are called the simple machines. You will find them in a tool shop, in a kitchen, and in all sorts of other places. You use some of them every day, or you see someone else using them if you happen to be looking at the right time.

Suppose you want to pull a nail out of a board. You tug at it with your fingers but it won't budge. So you get a claw hammer. You hook the claws around the nail and you push down on the handle and up comes the nail. It's easy. It's easy because you used a lever. The lever seemed to give you more force. And it did. That's one of the things machines can do — they can increase force. You have seen levers used to lift heavy rocks and other heavy things. You could lift a piano if you had a long bar and a brick or two and arranged them as they are here in the picture.

Of course you realize that a machine like a lever cannot give you extra force just free of charge. If one end of a lever gives extra force, then the other end must be moved farther. The end of a hammer handle moves farther than the nail is lifted. If you were to lift the piano as in the picture, then you would have to push down the long end of the bar much farther than the piano is lifted up.

Just as a lever cannot give you extra force free of charge, neither can a pulley. But let's look at what a pulley can do for you and what it will charge for its service.

A pulley is generally set inside a frame called a "block." You may have heard people talk of a "block and tackle" — this is just some rope or cord and one or more pulleys hooked together. If you attach one movable block and tackle to a 2-pound weight, you can lift the weight with about 1 pound of pressure. The block and tackle has doubled your force. It's charge? Well, for every inch the

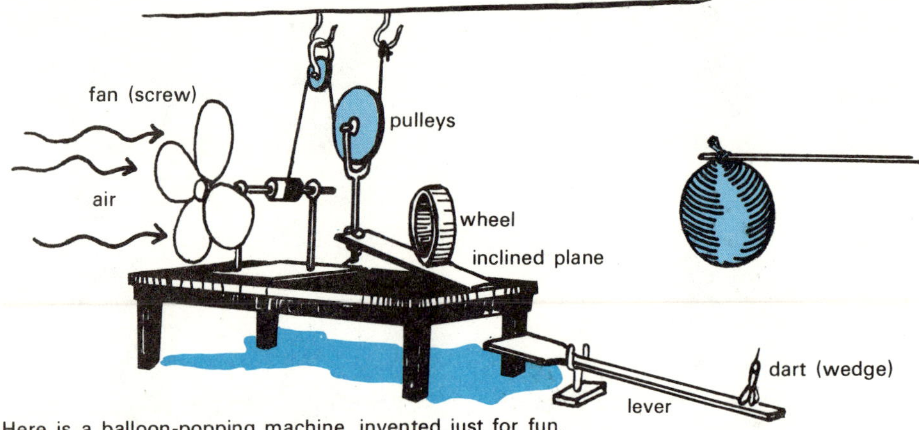

Here is a balloon-popping machine, invented just for fun. A draft of air against the fan will pop the balloon. Do you see how it works?